Anti-computing

Manchester University Press

Anti-computing

Dissent and the machine

Caroline Bassett

Manchester University Press

Copyright © Caroline Bassett 2021

The right of Caroline Bassett to be identified as the author of this work has been asserted by them in accordance with the Copyright, Designs and Patents Act 1988.

Published by Manchester University Press
Altrincham Street, Manchester M1 7JA

www.manchesteruniversitypress.co.uk

British Library Cataloguing-in-Publication Data
A catalogue record for this book is available from the British Library

ISBN   978 0 7190 8378 5   hardback

First published 2021

The publisher has no responsibility for the persistence or accuracy of URLs for any external or third-party internet websites referred to in this book, and does not guarantee that any content on such websites is, or will remain, accurate or appropriate.

Typeset
by New Best-set Typesetters Ltd

This book is for Rosie. I thought you might like it.

Contents

List of figures	*page* viii
Acknowledgements	ix
List of abbreviations	xi
1 Anti-computing: a provisional taxonomy	1
2 Discontinuous continuity: how anti-computing time-travels	34
3 A most political performance: treachery, the archive, and the database	69
4 No special pleading: Arendt, automation, and the cybercultural revolution	105
5 Polemical acts of rare extremism: Two Cultures and a hat	140
6 Apostasy in the temple of technology: ELIZA the more than mechanical therapist	168
7 Those in love with quantum filth: science fiction, singularity, and the flesh	186
Conclusion: Upping the anti: a distant reading of the contemporary moment	217
Index	243

Figures

1	An affective taxonomy of anti-computing	*page* 25
2	A taxonomy of theoretical positions	27
3	A general taxonomy of anti-computing	221

Acknowledgements

I wrote this book in three halves at least and it took me far too long. Life, work, and COVID intervened. I wouldn't have finished it now without threats and encouragements from close friends and colleagues, particularly Kate Lacey, who told me to stop, and Sarah Kember and Kate O'Riordan, who told me to keep going – it's true we wrote a different book together, but this one is finished because of that one.

This project was completed at Cambridge but it began at Sussex University, and I also worked on it during my time as the Sanomat Fellow at the hugely hospitable Helsinki Collegium for Advanced Studies (HCAS). Thanks are due to many at the international community of scholars at HCAS for feedback and for many inspiring discussions. Huge thanks are also due to my former colleagues in the School of Media, Film and Music (MFM) at Sussex. I am particularly indebted to the MFM Women's Reading Group for their forensic feedback. I'd also like to acknowledge a debt to Ben Roberts, whose thinking has influenced the discussions of 1960s cybernation debates and of contemporary automation anxiety in this book very directly.

Above all I want to thank past and present members of the Sussex Humanities Lab; in particular James Baker, David Berry, Kat Braybrooke, Alice Eldridge, Beatrice Fazi, Wes Goatley, Emma Harrison, Tim Hitchcock, Ben Jackson, Sally Jane Norman, Ben Roberts, Rachel Thomson, Amelia Wakeford, Sharon Webb and David Weir; but there were many more people there who were inspiring, cheering, or noisy in a good way. Modular synth and robot opera have nothing to do with the contents of this book, but the sanity of people working in those areas saved mine.

Some other acknowledgements are due; I interviewed Alice Mary Hilton in New York shortly before her death. Her recollections were not always comfortable, but they opened a door into an earlier moment of automation fever and began this project. I'd also like to acknowledge the Special Collections unit at the University of Sussex, where the Matusow Collection is housed, and to salute the infinitely patient Matthew Frost and others at Manchester University Press. Thanks to readers named and unnamed, but definitely including Jussi Parikka. Finally, thanks to Kate O'Riordan (again), Cecile Chevalier, Hilary Baker and Jane Bassett for solidarity above and beyond writing.

Abbreviations

AI	artificial intelligence
CP	Communist Party
CPUSA	US Communist Party
FBI	Federal Bureau of Investigation
HUAC	House Un-American Activities Committee
ISADPM	International Society for the Abolition of Data Processing Machines
IT	*International Times*
OHID	On Human Individual Dignity
SEFVA	South East Film and Video Archive (part of the Matusow Collection, Special Collections, University of Sussex)
SF	science fiction

Chapter 1

Anti-computing: a provisional taxonomy

Anti-computing punctuates computational advances; the rush ahead, the demand to be wanted, the claims for progress, and for progress as automatically good. Anti-computing is a pause, a stop, a refusal, an objection, a sense, an emotion, a response, a popular campaign, a letter, an essay, a code-work, a theorist, a sensibility, an ambience, an absolute hostility, a reasoned objection, a glitch, a hesitation, an ambient dislike. It may be articulated by a human, a crowd, a network, or by a program that refuses to run. It may also be an element of an assemblage containing many other elements; some of them in conflict.

Anti-computing takes many forms. It is unique to its moments of emergence, but is also characterized by recurrence. Many (though not all) of the components of anti-computing today are familiar even if they arise in response to a specific formation 'as if new', even if the claim is that the issues they address are graver than ever before, entailing qualitatively higher – even existential – stakes. Anti-computing may emerge as something felt and believed, something cultivated. It may be presented as a rationally adopted position or critique, a feeling (perhaps a yuck factor), or it may be an operation. It is articulated as an individual view, a collective one, as a one-off, and as a circulating and time-travelling discourse (operating synchronously and diachronically – albeit in the latter case in disrupted ways). As a human response to the global embedding of the computational it materializes in heterogeneous ways, but since humans, after all, are not separate from the machines with which they co-evolve, this materialization entails technologies as well as humans. It is indeed partly technological.

Anti-computing is all of these things at once because it is part of, and responds to, a formation that is itself pervasive, intrusive, ambient, emergent, installing, materially heterogeneous. I am talking, of course, of the computational, and of computational culture; about a history, about informational capitalism now, and about the computationally saturated and only partly foreseeable future. Computers have spread. They have spread further and faster than earlier human generations imagined possible – or that earlier generations of computer appeared to make feasible. There are over two billion computers in the world, and many billions more embedded control circuits. There will be 20 billion sensors in the world by the end of the 2020s, the decade we have already entered as I complete this book. The numbers become meaningless – though their sheer weight remains significant. The extent of the computational has broadened and deepened; networking the world, entering the body, reaching into Space.

In view of this relentless expansion, campaigns against computerization, anti-computing critiques, hostile interventions, writings 'against', in the academy and in popular culture, might be judged to have comprehensively failed. Certainly, if the measure of success is less not more computing, then they haven't *worked*. Anti-computing arises within horizons that have been – and still are – dominated by acceptance of the computational and its expansion, whether this acceptance is constituted as compliance, enthusiasm, indifference, acceptance, disavowal, or fatalism; all these have surfaced as responses over the seven decades or more since early computers first made an appearance in public, during which time computerization has been a live social issue. Anti-computing is, viewed in relation to this massive and apparently unstoppable advent, and in view of mass acceptance of it, an exceptional response to computerization, relatively rare, relatively rarely consistently expressed, even more rarely persisting, or becoming an organized movement. Of course, since the late 2010s and now in the 2020s, there has been an efflorescence of commentary – governmental, cultural, social, academic – expressing or exposing hostility and concern about the digital. Arguably we are in a renewed age of amplified automation anxiety (Bassett and Roberts, 2020). Consider the endlessly multiplying channels across which this anxiety runs the (digital) production processes that make it (increasingly digital), or simply consider the growth in the number of sensors in the world since, say, Jaron Lanier's 2010 *You are not a Gadget*, which might have been an automation anxiety outlier and the latest internet scare book to be up-marketed on Amazon.

Given the above, it is legitimate to wonder whether critical thinking, writings, campaigns, or interventions of other kinds 'against computing' matter. Or how they matter. For instance, do they make a significant intervention, are they significant historically, do they persist; how do earlier forms of anti-computing connect with contemporary anti-computational turns? Or, to introduce another register, do they suggest ways in which a different attitude towards a techno-socially saturated future and present might be cultivated? Could this be one which responds to a situation in which computational capitalism has reached the limits of viable expansion (environmentally) even as its growth continues? This is one of the ways the world is being broken, and one of the ways we are not repairing it. What do different forms of hostility or refusal suggest about the terms of the exchange or interaction between computers and (other) material and immaterial elements: bodies, discourses, texts, and (other) machines, intelligent or not; and what kind of alternative propositions can they open up – even in, or by way of, their intransigent refusal to conform?

The wager of this book is that historically arising and contemporary anti-computing formations matter and have more to do with each other than might at first appear. One example: contemporary surveillance concerns and hostility to comprehensive data capture by states are often explained as arising because of new data-capture techniques (big data), but they are also partly shaped by, and certainly resonate with, at least thirty years of fears of a data society. We have figurations and imaginaries to reach for when we discuss data surveillance, even as we declare it all new when we encounter it in a specific form, or at a specific moment, as a crisis not yet narrativized and thereby contained, let alone simply turned into information (Doane, 1990). You might say we reach for old stories, but we also need to ask how these stories materialize in new ways.

How, then, is anti-computing to be investigated? In particular, how are anti-computing moments of the past to be identified, given the thoroughness with which the relentless march of computerization obliterates traces of earlier resistance, hesitancy, and hostility in its rush to claim the present, and given its designs on, its foreclosure of, the future? Furthermore, once 'found', how are such formations to be explored? Addressing this last question demands not only identifying traces of the anti-computational, it also provokes a reconsideration of how the materials and discourses involved in its formation and articulation may be theorized and understood in relation to specific or located instantiations, and more broadly.

A dominant – teleological – understanding of 'what technology does to society' doesn't help to address these questions. It buries them. Computerization is often visualized as an inexorable progression. This progression, often taken to be synonymous with human progress, or understood as its leading edge, attracts support; at its most ardent this becomes quasi eschatological – something like the hope and desire for new gods, or a new regime of power to solve the problems of an old order. The latter is a sensibility evident in strong versions of singularity discourse, which see in artificial intelligence (AI) the chance to generate new kinds of post-human super-being. But a sense of the inexorability of computational growth also infuses visions of the coming quotidian, and is widely offered as a matter of fact, the apparently given and unarguable reality of the present and of the future. Computers, it is widely argued, are not only shaping life today but are increasingly also dictating the way the future will go. Granted, it may also be admitted by those who wish to follow it that this path is not entirely foreseeable. The complexity of interwoven processes of innovation, and perhaps also the much-heralded arrival of new forms of generative complexity, notably through advancements in machine learning, mean the logics of computerization are not fully possible to discern (at least by humans – and there is still nobody else). Despite this, however, the trajectory of much commentary on digital futures suggests a belief that – unknowable though they are – what *is* known is that these imprevisible logics will out. This belief, which is essentially teleological, turns any obstacles to the further advancement of computing – whether these concern narrowly technical issues, bottlenecks, reverse salients, or social tools including legislation – into temporary impediments to what *will* come, obstructions that will be routed around, sooner or later. Viewed from this perspective anti-computing interventions too, whether they come as intellectual broadsides, or emerge as circulating hostile discourses amongst various publics or as aesthetic or literary interventions, will only ever momentarily stall the disruptive energy of change, the forward march. They are, that is, deemed to be doomed to eventual failure or terminal subsidence. What are raised as problems concerning digital automation/instantiation – digital culture, AI for instance – may be taken seriously whilst maintaining this general perspective, but they are nonetheless also understood as proximate problems, as problems that 'we' may expect *to overcome in time.*

Consider the very different responses to the computational found in Joseph Weizenbaum's critique of the logicality/rationality equation, in Hannah Arendt's attack on the cybercultural idealists of mid-1960s New York, in F.R. Leavis' attack on technologico-Benthamism, in anti-database elements of the US and UK counter-cultures (all discussed here), or consider the more recent writings of Morozov (2012), Pariser (2012), Lanier (2010), Noble (2018), Hawking (2015), respectively railing against the cloud, the crowd, algorithmically accelerated racial bias, and the rise of computer intelligence. All partake of something I am terming the anti-computational. All are – if the dominant logics of computational progress are accepted – sideshows to the main event. They have been or will be absorbed, set aside, sidelined, routed around, dismissed, put in their proper place. It isn't an accident that futile or mindless resistance is the popular (and distorted) meaning currently widely ascribed to Luddism,[1] or that the term Luddite is now invoked far more often in relation to computational technologies than to its original contexts.[2] Perhaps this is why an avowed technological Luddism, proffered as a form of computational dissent, never had much purchase, though it was demanded in the early years of the public internet (see e.g. Webster and Robins, 1986) and is undergoing something of a recurrence (e.g. Lachney and Dotson, 2018).

A sense of the inexorable expansion of the computational is easily discerned as long standing, part of a structure of feeling emerging alongside the digital as it becomes pervasive in 20th- and 21st-century societies (albeit unevenly distributed). Consider the 'drift' (in fact a powerfully flowing current) towards the adoption by nation-states of bulk surveillance techniques no longer based on 'need to know' rationales but on exhaustive capture, which may be disliked by many but which are not being widely or systematically contested. Surveillance has already become embedded, an intrinsic part of the wider social architecture (and not only in the West), including the architecture of the built environment and of the self (see e.g. Andrejevic, 2015). Infra-structuralization produces a lockdown, not only cementing technological systems but also constraining the cultural imaginaries that might let us imagine them operating in other ways, let alone not operating at all. The teleological horizon, then, in the end acceding to a form of fatalism, leaves little space for the impertinent intrusion of the anti-computational. Anti-computing, whether understood as worthy or foolish or tragic, hubristic or pathetic, ill conceived, optimistic, or even heroic, is within this horizon ultimately always to be understood as a *futile* form of resistance, at least if it intends to do more than rectify an immediately correctable, temporally discrete anomaly.

In response, the wager of this book is that teleological readings of the computational, both those found in popular culture and those generated by particular framings of the technological, can be disputed, and that accounts that fail to register the significance of dissent and hostility can also be challenged – and should be. The anti-computational favours a dissonant reading of computational history, demands an alternative understanding of the technological, and can generate an alternative account of the actual and potential impacts of the computational. It does so as an act of writing, but also as an act of excavation.

Acknowledging that anti-computing moments falter, and that anti-computing often fails, and that there are many positive aspects to the computational, I set out to acknowledge and explore dissent, hostility, antagonism, doubt, unease, in many forms. Teleological accounts of computational technologies and the computerization of culture, taking various explicit forms, working in multiple registers, silently informing many accounts, *need* to be challenged. Both because they operate in powerful ideological ways and because (partly because of this) they tell only half of a larger story; anti-computational formations are an element of, and an element essential to the understanding of, those processes which Bernard Stiegler memorably defined in terms of a co-evolution between humans and technology and explored in relation to medium technological times in terms of grammatization or exteriorization and its potentials for good or ill (Stiegler, 2013).

These are times in which assumptions of unlimited space for growth or progress, or of humans' ultimate mastery of our world, are revealed – in the flames of Australia, or in the melting of the ice, or in COVID-19 – to be the fictions that they always were. A theory of technical co-evolution, challenging assumptions of mastery and control by 'the human' over 'nature' by challenging the boundaries between them, specifically challenges technologically enabled visions of automatic or open-ended progress. The engagements between the computational and the environmental are complex and multi-layered, but also brutally obvious; land poisoned by tech manufacturing processes, the rise of throwaway culture, labour exploitation around rare metal mining for smartphone components, pollution, the energy consumption of data lakes used in AI. These issues are explored elsewhere (see Solnit, 2007, 2013, Bauman, 2003, for early engagement, and emerging writing in ecological digital humanities for more contemporary treatments, including Yusoff, 2018, and Smith, 2011) and do not directly constitute the proximate subject matter of this book. However, the broad attack on the presumption that unlimited growth represents progress (and perhaps virtue), made for instance by Latour and Haraway in discussions of technology and environmental limits, resonates with many explicit critiques of the computational as the handmaiden of such forms of progress, and is present in a more amorphous form more widely (see Latour, 2011, Haraway, 2016). Work by Timnit Gebru et al. on the high environmental/energy cost of expanding data lakes used in AI sharpens the critique in relation to emerging technologies (Gebru, 2021).

Significance

The anti-computing formations I explore here are significant first of all *because their very existence* provides evidence that in earlier decades the path 'forwards' towards greater computerization has not been as direct as it appears to us in retrospect. Foucault (1972: 14) talked of the 'tranquilized sleep' of a conventional history and asked how this might be disturbed. Anti-computing formations punctuate those breathless but still somnolent accounts in which the 'progress' of the computerization of culture, viewed as the latest/last work of humans, is assured through the smooth and relentless application of (computer) science. An exploration going by way of the anti-computational exploits this disruption and theorizes its significance. Anti-computing can be understood in this respect as a partly theoretical construction or methodology, one which isn't quite archaeological in Foucault's sense, nor quite a media archaeology, but which shares with both of these some interests in discontinuity, series, scales, irreducibility, and looped connection, seeking out and exploiting the possibilities opened by interruption – and systematizing them where possible.

Anti-computing exposes other ways of seeing the history of computational adoption and expansion, and discerns ratios operating between humans and machines other than those most conventionally remarked upon. Of course, there are multiple ways in which dominant forms of computational culture – computational capitalism and its growth logics – are questioned. The anti-computational moment relates to other forms of resistance, for instance those that can be framed as *doing* otherwise (hacking 'fixes' to reopen apparently finished technologies, the adoption of alternative modes of use, turning data against its owners, all are good examples here and are also much studied; see e.g. Jordan, 2017). Anti-computing makes its point differently. Its modus operandi is dissent. It tends to prioritize refusal and critique over appropriation. This matters partly because, as is increasingly clear in relation to commodity culture and culture jamming, appropriation only too easily produces *reappropriation* – net autonomy was always going to be temporary, as Hakim Bey (1991) put it long ago. Dissent may (also) jam up the technological imaginary, but this time through forms of distancing, refusal, doubt, rejection. Anti-computing is not about playing with, but rather about *not* playing at all, or at the very least about questioning the logics of the game.

A certain irritation might be legitimate at this point; what *is* anti-computing? So far, I have defined it as a series of found and generated responses to the computational that wish to question some of the powerful assumptions about the inevitability of computational technology and computational growth and/as progress. I have suggested that they may be powerful in themselves, and that their existence disrupts a story that is all too easy to tell, that threatens to tell itself. I have also begun to talk about anti-computing as an excavatory methodology. I recognize that these two definitions provide only the beginnings of an answer, and they are intended to be partial. In this book the term is allowed – indeed *encouraged* – to morph, to expand and contract, and to change case. Indeed it is more than a term, although the matter of terms is where I turn to next.

Computing and anti-computing

To explore anti-computing demands some discussion of terms such as 'computing', 'the computational', and 'computational culture', given the multiple ways in which they are deployed. A computer is a machine that operates on data according to a set of instructions (an algorithm). Digital computing is a particular instantiation of such a machine; this distinction can be clearly made – but it needs to be recognized that the two terms are endlessly switched in popular culture. The computational is not the cybernetic, certainly not as this was defined by Norbert Wiener (1950), but some argue that the latter constitutes something like the basic architecture of the former. Competing definitions are common; computing 'began' with Babbage, or with Lovelace and programming, or with Hilbert's work on algorithms, or with Turing, or with the Turing Machine, or with Turing's paper 'Computable Numbers' (Fazi, 2018). Or, in deep media accounts, it began centuries sooner (Zielinksi, 2008). These definitions inform more or less contemporary analyses of computerization processes in society and culture; consider Wing on computational thinking (2006), Lanier on the computational (2010), and Berry on the computational turn (2011).

I employ some of these definitions here, chiefly using them in the following ways. First, computing is taken to mean something reasonably close to Turing's original definition of the universal machine, capable of calculating according to coded instructions (Turing, 1950). This draws attention to computing as a process and an operation which may be instantiated in many ways; on this basis the digital (digital computing) is a particular kind of computing, quantum computing another. It also allows that computational operations may be components of assemblages taking many physical forms; this is highly pertinent today as billions of embedded computer controllers animate and endlessly expand the emerging Internet of Things (itself a term now approaching senescence).

Second, computing, the computational, computers: all these are also allowed to be umbrella terms, designating assemblages widely understood to be computers, and/or operations understood to be computational at the time of *this* writing, or in *their* time. Today these include that expanded constellation of personal computers (PCs), tablets, cell phones, and cloud storage devices and facilities that constitutes the network central to computer culture: the internet and the platforms. To this are added those things and processes that are computerized (becoming computer controlled, or 'becoming' computational operations) and that are, in the process, transformed, often via a process that might be termed 'sensorship'. The qualifying criterion for what constitutes a computational object or process isn't absolute and is contingent. In 2020 prototype driverless cars are often understood as computerized vehicles; their capacity to automate what has been viewed as a form of human manual and cognitive expertise – driving – is what marks them out. But digital circuitry is of course deeply embedded in all reasonably modern cars (even mine, which still has cassette technology and mechanical windows), even though these are not conventionally recognized as (defined as) computational objects. The sociology of media-technological innovation, with its exploration of the becoming invisible of technology through appropriation, which produces a culturally informed sense of the technical per se, has much to teach here (see e.g. Silverstone, 1994, Hartmann, 2009). Similar ambiguities arise elsewhere. Consider smart homes technologies. Fridges transformed through their computerization (embedded sensors, storage, control circuitry) into informational and media devices continue to perform their traditional chilling and storing functions. But what is milk to me is now information to a supermarket. The computational is a relational category.

Despite the flexibility of these terms computing does *mean* something, not only as a technical definition, indicating an assemblage with the capacity to undertake particular operations, but also in the shifting vernaculars of the everyday. Allowing for this I bring into my orbit discussions that, despite the vagueness with which they may adumbrate specific objects, or specific computational technologies, nonetheless focus hostility onto computing, or the computational moment, however they define it. This focus is sometimes a part of a larger critique: of technology in general, singularity in general, or modernity in general, for instance, but it is there. This focus shapes my own inquiries. Later in the book, for instance, I explore Arendt's treatment of automation, beginning not with *The Human Condition* (1998 [1958]), a work of philosophy exploring the *viva activa*, but by asking how Arendt fared among the cyber-enthusiasts and haters, industrial organizers, civil rights activists, and nascent tech industry professionals who argued with her face to face at a conference exploring the 'threat' (or promise) of new waves of specifically computational automation or 'cybernation' in the early 1960s.

One more definitional comment. I am wary of using 'the computational' as a term to replace 'the digital', at least if this switch is taken to mark an exact break point, since I don't believe it does. Terms such as computational, digital, postdigital are invoked as periodizing categories in cultural or critical theory and do have useful heft as aesthetic provocations (on the postdigital see Florian Cramer, 2014). However, their usage isn't precise. In the case of the digital and the postdigital, for instance, there is obvious slippage across registers marking the aesthetic and the sociological – and that produces the peril of an unexamined assumption that we're all in this together, and all in the same place. We should ask of the postdigital, post for whom? post where? post when? – and recognize that the response 'it's post for me!' often isn't good enough. Not least because it fails to recognize (post)colonial issues of historical discrimination and normative legibility. In another context the scholar and artist Lewis R. Gordan has discussed these kinds of universalizing labels in terms of the epistemic closures they produce, particularly around race (Gordan, cited in Evans, 2019), and this is apt.

The computational, then, is understood as a formation, one in which computational control logics dominate, but one which invokes and assembles many other actors in operation. The computational cultures that arise as the logics of computational assemblages expand through becoming operational are specific to particular times and places, which is to say they operate in relation to, and as an integral part of, specific political economies. Computational culture today is not necessarily isomorphic with neoliberalism as a political economy and a social order, although the computational is a key mode through which neoliberalism is organized and operates – and is also important in how it sees itself (as an information society, knowledge society, or, in so far as the label 'computational capitalism' is accepted, as a computational society).

Anti-computing instances

In this book anti-computing is explored from two directions. First via a series of investigations of anti-computing in specific sites and contexts, scattered across the seventy years or so of post-war computing. In each, the goal is to understand anti-computing in a particular social and material context, to ask what forms of understanding underpinned anti-computational interventions (how the technological was constituted as powerful, how it was understood in relation to political economic conditions pertaining, for instance, or what justifies my labelling of it as anti-computational). These investigations focus on some (of many) instances where aspects or instantiations of the forward march of the computational are attacked.

There are no pretensions to completeness here. They would be frustrated, since anti-computing is found in many places, takes multiple forms, and operates at different scales. Consider these various iterations: campaigns against computer surveillance, for instance early opposition to the Regulation of Investigatory Powers Act 2000 in the UK, were and are anti-computational, as were objections to the potential data capture operations of a single software package, the finance package Lotus 1–2–3 (an early example of a tightly defined anti-computational moment). Then there were the bundles of assembled new technologies that promised to 'cybernate' various forms of work in the 1960s, or the electronic page make-up systems central to shifts in the print industry in the UK in the 1980s; both aroused a hostility that was anti-computational. The UK print case is a useful example, since it illustrates how the anti-computational can draw itself up into and articulate antagonisms far exceeding the purely technological; the print industry wars were part of the Thatcherite attack on trade unions and working-class organization in the UK in the 1980s (Cockburn, 1992). The computational informs cybernetic routes to the post-human – and so warnings against these forms of post-human existence may usefully be explored in anti-computational terms. Or – bathetically perhaps – consider those on trains who prefer silence to electronic noise. Or those who prefer thick, rich, 'real world' contact to virtual interaction, or those who take detox 'holidays' away from the saturated computationality of everyday life (see Harrison, 2020), or who call for organic community over techno-scientific culture, or who critique technocratic rationality as that which will hollow out the political world. All of these positions and arguments are examples of anti-computing. Some of them are explored further in this book.

Hopefully, these examples already suggest that there is something impure about the anti-computational, something incommensurate about the various scales at which it operates, and something partial and rarely complete. This impurity interests me and is partly why I have not set out to research the most obvious cases. I am interested in interventions, often now largely forgotten, aired in very different sites, all of which took forms that, even if they were staged by academics, reached beyond the spheres of the academy as these latter are narrowly conceived of, and certainly beyond critical theory as an entirely contained project sufficient unto itself. Rather, anti-computing constellations, which entail an engagement with theory, are considered as they emerge in real-world contexts. The attempt is to produce critically informed rich descriptions. A series of events, writings, moments, that cross between theoretical and everyday registers, and that are not often explored for what they say about the computational, are reopened and explored *as* anti-computational, and explored *through* the optic this taxonomic division provides. This close-up view, somewhat set aside in the rest of this opening chapter, or at least not referred to in specific terms, constitutes the bulk of the work undertaken.

Anti-computing and its travels

The second concern of this book is to explore what connects apparently discrete moments of anti-computing that irrupt in response to the new. It should already be clear that the objects, assemblages, movements, writings, that constitute anti-computing are markedly heterogeneous, a term that, as Fredric Jameson notes, has had spectacular success in late capitalism (Jameson, 2015). Even in this heterogeneity, however, specific iterations of the anti-computational are often regarded as disjunct from one another and their connection to earlier forms often not recognized; it tends to be 'new' dangers, 'new' forms of sociality, 'new' kinds of addiction, 'new' kinds of AI that are feared or disliked, after all. This suggests that discontinuity wins out. On the other hand, there are also connections to be identified and explored. Anti-computing is multifaceted and diverse in its materiality, orientation, and scale, unexpected in its forms, unique in its instantiations, but it does exhibit forms of coherence, and exhibits traits held in common. Moreover, although the *times* of its emergence cannot be predicted, that forms of anti-computing *will* emerge and re-emerge over time can be anticipated, at the very least.

Thus, whilst discrete forms and moments of anti-computing are explored for their intrinsic interest, the question also tackled here is how to understand anti-computing as a formation that regenerates, takes up, takes on, moves on, and revives in various *recurring* forms across time, in irregular but nonetheless recurring circuits. How do these forms travel, how do they translate, come to be rematerialized? Does anti-computing itself, as a formation, have a form of coherence over time? Is there a way of doing technological history that can encourage recognition of this coherence? This last begins to ask questions about technology and history, and will return us, in Chapter 2, to an engagement with media archaeology. Briefly prefiguring that here, my argument is that a conditional and partial continuity, complementing the inaugural moments of new forms of computational dissent and confounding what might have appeared to be the terminal disappearance of some others, constitutes one of the defining dynamics of anti-computing. The question, then, is how to identify and account for discontinuity/continuity, to explore what appears to be – and what is – durable in the relationship between computers and the (techno)cultures they intervene into.

This proposition, constituting the second focus of the book and supplementing the concentration on discrete moments, also begins to redefine the anti-computational itself, since the latter becomes recognizable in part as exhibiting a kind of discontinuous continuity. Here work by the historian Valerie Traub, seeking to understand historical recurrence as well as synchronic emergence, has been found insightful. Exploring the periodic rematerialization of various bodily tropes across long stretches of human history, Traub sets out to understand how various perennial logics and definitions remain useful, across time 'emerging at certain moments, silently disappearing from view, and then re-emerging as particularly relevant (or explosively volatile)'. As Traub sees it, such 'recurrent explanatory logics … are subject to change', but also 'evidence a form of persistence' (Traub, 2007: 126). Older tropes, apparently consigned to history, may come back to life to 'trespass' on the new. Anti-computing may be explored, at least initially, in terms of these dynamics. That is, it can be viewed as a series of perennial logics, operating within the horizon of the computational, or of computationally influenced times, and disrupting them. Within this time frame, which is massively compressed compared to the long historical periods with which Traub is concerned, forms of anti-computing emerge and find force and purchase through various modes of materialization, and then fade in 'relevance' before possibly reviving and returning. These oscillating circuits or, borrowing from Traub once again, these 'cycles of salience' (Traub, 2007: 126, 133), are irregular and partial, and new irruptions always contain the discontinuous, as it were. However, in the sense that the return is likely to come, the anti-computational moment can be characterized as erratically predictable.

An initial taxonomy of the anti-computational

Anti-computing exists and is found operating in the world; this is the burden of the preceding section. Anti-computing, as I define and use it, is however, not only a real-world formation but also a methodology, a working thesis, and an investigatory heuristic. It is to be developed as a critical methodology or a 'working concept' (Foucault, 1972: 9) and the taxonomic operations set out below begin that task. They suggest commonalities, isolate matters of interest, and expose or suggest connections across time, location, and register. The task is to identify a series of initial taxonomies, identifying contingently stable and recurring logics or tropes of anti-computing.

Consistent with this, these categories have been constructed in response to what is found 'in the wild' (how popular culture might group various elements), in response to what my investigations of particular anti-computing moments across the decades suggest, and because particular principles of grouping might offer insight into how the formations under investigation can be understood. What results – some possible working taxonomies – are then themselves set to work, the taxonomic being recognized as a mode of categorization that is both descriptive and operational, a suggestive model with some performative force, rather than anything that will be fully realized. It becomes a means through which to consider rising forms of anti-computing, to ask how they are continuous and discontinuous with what has come before (as is further explored in the final chapter).

Polarities/binaries

Taxonomies may be built, or found in operation. Either way, they imply the finding or the identification of a pre-existing order – but are themselves a powerful ordering process, constituting a mode of interested (non-innocent) knowledge production. As Foucault famously taught us, or, rather, Borges did, social and other taxonomies may tame the wild profusion of things by the imposition of categories – which may be apparently incommensurate with each other, or which may overlap in various ways (Foucault, 1972, Borges, 2000). Moreover, the designation of categories may become the subject of contestation, or be part of a political struggle, a demand, as Foucault put in an article exploring critique, not to be governed 'like that' (Foucault, 1997).

Building an anti-computing taxonomy, it seems logical to begin by identifying a naturalized or dominant taxonomy of anti-computing, informing much debate around the computational in public culture but also found in theoretical writing, the chief characteristic of which is a blunt binary division. This organizes hostility to the computational into two categories, dividing hostility to computers arising 'because computers are particular kinds of machines' from hostility arising 'because human societies are ordered in particular kinds of ways'. The first category concerns computer power, and the second the amplification of systems of human domination through the use of control technologies.

The first category is essentially ontological. Something intrinsic to computers, the computational in essence, taken variously as instrumental, indifferent, in-organic, in control, or simply as 'alien' (see Weizenbaum, 1976), provokes fears, anxieties, hostilities, and dissent. These fears expand as computational cultures themselves expand into new zones. The invocation of a familiar meta-discourse from within this category constructs or produces an equally familiar machine; this is 'the computer' that stands against 'us' humans because it threatens to control us. It is why, as the technologist Bill Joy put it about AI back in the 1990s, we should be very afraid of the future (Joy, 2000); the same message is now being repeated around AI singularity issues.

The second category, the 'wrong hands' category, is centrally concerned with political economy and human determinants. Its central concern is with what those in power can do with computers, and who those in power will be, or are – by virtue of that character of the computational (what Winner called political technology). It fears that those who control computer systems use them to accelerate control over the 'rest of us'. This category attaches itself firmly to computing, but we also see earlier iterations in other media technological forms. Consider Orwell's *1984*, where the evil media was television, or rather the all-seeing telescreen zooming state violence – and, by the way, fitness classes – into the homes of citizens.

Working with this categorical division, two distinctly different ways to account for the recurrence of forms of the anti-computational arise. In the first case a certain ontological consistency is presumed – computers continue to provoke a particular kind of hostile response because of what they are and continue to be. The second category opens the way to account for the recurrence of certain forms of hostility to emerging forms of the computational in social terms – for instance, via the argument that the society that has produced computers is itself, in a fundamental way, static, operating with consistent social logics or taking the same structural forms; that the same threats and problems with the computational, which are really 'problems with society', continue to irrupt is, then, not surprising.

This polarized tale, this binary division between computer ontology and social organization, has wide purchase. It emerges in popular discussions of technology and culture and finds echoes surprisingly often in techno-cultural theory and critique. It is powerful, operating as what Erkki Huhtamo (1997) might call a *topos*, or leading metaphor, in discussions of computer and society. Ultimately, however, these are false dichotomies and explanations that work with this model to account for the anti-computational. To push this somewhat, my claim is that sustaining absolute divisions between the ontological and the politically economic, here articulated as a division between what computers might do to 'us humans' by virtue of their ontology, and what (bad) uses humans (or human societal organizations peculiarly stripped of their material supports) might put computers to, is to support the unsustainable – in theory (where it produces a distorted account) and in practice or operation (where it operates ideologically).

To ask the (apparently) social question 'What it is about the social systems within which computers have arisen that provokes the return of hostile responses to the computational?' is immediately to provoke questions about the distinction between technology and society, to ask how the constitutive role of the computational within a social system can be understood. Langdon Winner's sense of technology as a political actor (Winner, 1985) would be the critical correlative here. The powerful ideological force of this binary taxonomic division should not be set aside. It is what lets actors such as Mark Zuckerberg claim it's culture, not technology, that produces platform violence. It is also what lets governments that have instantiated anti-Islamic policies and encouraged racial hatred insist that it's the (social media platform) technology, not the governmental policies, the social media gurus, nor the politicians, who are at fault for fundamentalism (which is not to say they are innocent either). However, I also want to insist that even as this division is found in practice, it is also in practice where this taxonomy can be seen to break down. There is rarely, if ever, a purely ontological basis upon which hostility to the computational arises, and no purely social basis either, once it is accepted that the social is techno-social. The interaction between materials and their constitution, and social structures and their operation in specific historical and located contexts, is misrecognized in this binary taxonomy, its purchase notwithstanding. Anti-computing, we might say, always has coeval motivations. Elaborating a useful taxonomy of anti-computing thus means recognizing *as* extant, but also rejecting *as* a naturalized division, and therefore rejecting *as* an appropriate tool for analysis, that binary division between categories of anti-computing that divide political and ontological factors.

Providing a better starting point to think about anti-computing and its characteristic forms demands keeping in play ontological and social factors of the computational, looking at durable and inaugural elements of each, and exploring their tensioned relationship in operation. For instance, to understand the relationship between/ difference between hostility to database expansion in the UK in the 1970s, in the late 1990s, and hostility today, two sets of continuities and disjunctions need to be taken into account. Computation took very different forms at each of these periods, but material consistencies, the digital in this case, remain (this might be a matter of computational ontology); this is the first disjunction and continuity set, as it were. Second, what is required to build a useful taxonomy is *both* recognition of the specificity of the political climate of any relevant period in computing's social history *and* acknowledgement of underlying consistencies within the social order across this period – capitalism persists, albeit it in remodelled forms, and its logics are dominant. The principles informing this example are now taken up and contribute to the elaboration of a new taxonomy of anti-computing which sets out to crosscut and disrupt the binary logics of the first taxonomy whilst also recognizing its framing power.

A general taxonomy: eight forms of anti-computing

The new ordering now set out is composed of eight categories, constituting a provisional taxonomy of the anti-computational, an anti-computational catalogue. It is generated through consideration of extant instances and/or recurring examples of anti-computing across some decades. New forms of the anti-computational may find their place in this general order, deform or stretch it and thereby remake it, or constitute a global challenge to it.

Anti-computing (i): computer technology as control technology

Computers are control technologies. This category responds to that. It concerns itself both with computer autonomy and threats to human autonomy. Asking 'who is in control – humans or machines?', it fears that the answer to this is that computers themselves are 'out of control'. But it also includes, and on an equal footing, concerns around the social order and questions of social control, power, and domination. Merging ontology and political binaries, it encompasses existential *and* political concerns. It fears the capacity of computers to deliver totalizing power to the state and to the market, cementing existing authority and power by ceding control of computational resources and foreclosing the open horizons of the future. Left critiques of technocratic rationality come into this category, for instance those reaching back to the later Frankfurt School (Marcuse in particular), but so do conservative variants of anti-computing, those who fear 'the computational' as that which will undermine the existing order of things where that order is regarded as just, fair, or simply desirable according to the situated position of those defending it. Hostility to groups who assert their counter-authority and threaten 'law and order' through computer use and appropriation often evidence this kind of anti-computing impulse.

Anti-computing (ii): computers becoming more lively

This form of anti-computing responds to fears concerning the displacement, and ultimately the *replacement*, of 'the human' by 'the machine'. Central concerns include the rise of computer intelligence and the extension of various forms of computer-delivered automation. The existential variant of this category fears that humans have no place in the future, since the advent of advanced forms of AI will mean the replacement of humans, either by their machinic successors or by future humans sufficiently different from us to be unrecognizable *as* humans. A less elevated set of concerns and anxieties, also coming under this heading, coalesce around issues of replacement as they pertain to labour and work: white-collar unemployment through augmented intelligence of computers, service sector and other labour replaced through the extended reach of computers into areas traditionally regarded as human specialisms because they entail particular forms of emotional intelligence or human imagination are key – home robots taking on care, teaching robots undertaking education, for instance. Many of these concerns are captured as fears around the automation of expertise or the replacing of human expertise with computational expertise in new areas. Anxieties around the displacement of the body, in relation to the bio-digital – for instance, the ever-closer coupling between humans and their devices (partial cyborgization) – can also be fitted in here.

Anti-computing (iii): computerization and the hollowing-out of everyday life and social interaction

This category captures anxieties that arise around the computational transformation of the everyday world, of social interactions, and of everyday life and the ways in which it can be lived. They focus on the surveilled self and life, platform sociality, the hollowing-out of social interactions through their automation, felt perhaps in spatial and temporal terms (slow movements respond to this phenomenological threat), and on the loss of bodily richness and the pleasures of the informing sensibility of a located physical embodiment. They also concern the lack of depth of new forms of everyday life that distributed and virtualized interactions are felt to produce. Concerns about the attenuation of attention – the continuously disrupted life – also fit in here. Latterly, digital detox camps and other refusal programmes respond directly to these concerns.

Anti-computing (iv): computer technologies and the threat to human culture

Issues captured and grouped here include hostility concerning the rise of computational culture and the perceived threat to older cultural forms: narrative versus database structures, the logic of code versus the emotional intelligence of the human, the game over the story, graphic principles over aesthetics, data visualization over human art production, virtualization over materialization, mobile screens versus cinematic projection. One strong version of this says computerization debases culture even as/or even if it expands the latter's sphere of action/operation, or that it enables the completion of the project of technocratic rationality delivered as a whole way of life (the computational culture industries). Adding an 's' here to pluralize human *cultures* and recognize difference is also important. Doing so identifies a related but distinctly different set of concerns: that the computational reinvigorates a form of binary thinking and calculation that standardizes and universalizes, so that the threats responded to here are of computer-assisted standardization, conformity, and non-situated universalism. Concerns around AI and bias find a place here. Finally, there is a strand of anxiety around computational engagements with the humanities and the arts which fears the replacement of human creativity with computational instrumentality.

Anti-computing (v): the general accident/catastrophe theory

This category groups concerns that neither computers nor humans control the outcomes of increasingly complex processes that computers undertake, particularly in relation to biotech but also in relation to network complexity, neural networks, emergence, AI. The fear is that these developments make the more or less accidental emergence of catastrophic outcomes drastically at odds with the programs' (or programmers') original intentions more likely to come about. Relating to cultural theorist Paul Virilio's (1997) theory of the 'general accident' and Beck's (1992) discussion of the 'risk society', this kind of anti-computing often emerges as a response to genomics and environmental issues, but also in relation to neural networks, AI agency, and algorithmic bias. It is invoked in more partial ways in many other spheres – program trading and market crashes are good examples. Once again, existential fears emerge but here tend to focus on the fate of the environment rather than, or as well as, foregrounding the future of the human.

Anti-computing (vi): horrible humans

This form of anti-computing doesn't 'blame' computers for the wrongs it discerns. Rather, it excoriates humans for taking advantage of what computers increasingly enable humans to 'get away with'. Essentially illiberal, it says that the anti-social grounds of the machine, the distancing from real accountability that pervasive networks and the platforms provide, undermine collective moral responsibility and social norms that other social forms policed. The computer is not only a mirror to ourselves through which we see ourselves clearly and do not like what we see, it also enables or 'encourages' forms of action that humans would not previously have undertaken; 'because they are there', 'because I can', 'because I can't be seen'. Computers are disliked because they come close to enabling (what is viewed as) a nascent human degeneracy.

Anti-computing (vii): standardization/quantification

This gathers together a cluster of hostilities and anxieties coalescing around epistemology. Computer operations are framed as determinate and reductive, and it is feared their expansion will produce a world organized according to 'alien' logics. This includes concerns that computing will render the world into quantifiable chunks so that it may be indifferently operated upon. 'Solutions' will triumph over 'theory' in the sciences and the arts (Anderson, 2008), quantification over qualitative judgement and hermeneutics, and the imposition of standardization, uniformity, and the ironing out of exception will triumph over variation, difference, and human judgement. This category operates in relation to, for instance, forms of knowledge production, computationally derived managerialism, bureaucratically led thinking. A criterion focused on knowledge issues, it also entails a concern about human ethics and their overruling or bypassing.

Anti-computing (viii): too much information

This category captures concerns over information overload, but also includes anxieties about the fetishization of information capture, the fetishism of 'facts', the overproduction of information – whether as data, text, image, and the prioritization of production and circulation over interpretation. This category of anti-computing is one of the most long standing. It finds new articulations around big data. Its obvious connections to other categories make it clear that the divisions between these clusters are not fixed or impervious.

This initial taxonomy is informed by a close engagement with a series of forms of anti-computational thinking over time. However, it does not reflect a robust, data-driven investigation. It is incomplete and can be disputed. Some people I have shown this to have suggested another one or two (or more) categories – although no alternative mapping provided a more total 'solution', nor even a more complete map. It is necessarily provisional. It responds to different methodological approaches to material studied, it accepts the hybrid, it assumes crossovers will continue to occur, and that the power and force of vectors linking the various categories will change over time and may rewrite the whole. For these reasons, rather than despite them, it is a substantial improvement over the binary division already discussed, and not only because each category can encompass ontologically directed criticisms and those that relate to instantiated technologies within specific political economies. It enables an exploration of anti-computing, its temporality, character, rarity, and the forms it takes across time, that begins by assuming a complex relationship between the ontological properties of the technologies entailed in computational intervention and the horizons of social power and particular forms of social system (capitalism, late capitalism, neoliberalism) within which these operate, of which they are part.

The emotional register?

There are other registers that could be used to further elaborate a taxonomy of anti-computing, perhaps to produce a taxonomic series. The sort could be by results (effective or ineffective campaigns), genre (theoretical writing, public campaign, popular or public opinion, legal response), material (documents, campaigns, technologies), or along various time lines that might impose forms of periodization (anti-computing before personal computers, without wires, including mobile devices, after 9/11, before mini-computing, after the dot.com crash, since Web 2.0, and so forth. And there is always the emperor's robot dog …). I am not pursuing most of these here, in part because they tend to divide where I wish to join – for instance, I want to trespass between, rather than to demarcate, critical theoretical and other forms of anti-computing, and I am interested in how hostility travels through rather than *defines* an era, or is defined *by* it. There are, however, some alternative taxonomies that intersect in productive ways with the one set out above. The first, set out here in Figure 1, albeit in skeletal form, organizes anti-computing by emotional register.

This ordering crosscuts the earlier eight-part taxonomic catalogue, as is evidenced in specific ways in some of the later chapters – for instance, in relation to revelation, revisionism, and expert witnessing in the case of a computer science insider who becomes a hostile witness (Chapter 6); anger in the case of F.R. Leavis, the unlikely subject of Chapter 5; or in relation to joy when the protagonist of

Anti-computing: a provisional taxonomy 25

Acknowledged fatalism
Anger
Anxiety
Boredom
Censoriousness
Concern
Critique
Disillusion
Fear
Frustration
Hatred
Hostility
Refusal
Revolt
Unease
Revelation
Jouissance/joy

Figure 1 An affective taxonomy of anti-computing

a singularity novel celebrates embodiment over AI minds in Chapter 7. This list is suggestive, and again certainly not complete; perhaps no emotional mapping can be otherwise. It is also potentially misleading – at least, if it suggests that anti-computing can be comprehended primarily as an individually felt emotional response or as an individual response per se. What I am trying to capture, even when working with or through individual responses or stories, are forms of collective and ambient unease, the collective sensibilities, perhaps the minor-key structures of feeling, often as they are gathered up and expressed through a single example, or by an individual, that constitute anti-computing as a material social formation, articulated on multiple platforms and through many materials, within a dominant culture that is generally positive or accepting, or fatalistic about computational 'advance'.

It's clear that an obverse list of categories, one that does not work through emotions but, rather, deals with categories concerning reason, might also be generated; anti-computing may be categorized variously as rational, thought through, analytic, deliberative, irrational, for instance. These cuts, however, rather easily fall into binaries (rational/affective, reason/emotion), again tending to stress division where I wish to notice joins.

A taxonomy of anti-computing computing?

A further alternative or complementary taxonomy could be built specifically around (or perhaps could be restricted to) materials. This would include those forms of *computing* (following N. Katherine Hayles and others and taking computing as undertaken both by humans and machines) that might be regarded as, or exploited to forward the goals of, anti-computing: glitches for instance, varieties of hacking, simple machine breaking, could be included in a broad spectrum anti-computational mapping. It would also be possible to categorize anti-computing in terms of the forms of technology or material or matter that 'resist' the computational (vinyl, shellac, paper, ink, for instance), or as a collection of forms of cultural acts or practices that in their materiality or form frustrate the computational; symbolic language with its ambiguity, the codes of the unspoken or unsaid, that which is only implied, glossolalia, face masks designed to disrupt face recognition, such as those developed by Zac Blas, for instance (Blas, 2012–14, Bassett, 2013). Some of these forms are discussed in later chapters. These groupings, whilst suggestive, are not pursued further here, but they do inform the whole. They might be seen to crosshatch the main grouping.

Categorizing critical approaches?

Anti-computing is heterogeneous in mode of address and form; notably, it is neither solely an intellectual formation nor solely a popular one. Various theoretical orientations, more or less directly engaging with anti-computing, might be permed out and captured under another set of headings (Figure 2).

These critical and theoretical orientations inform the eight-part taxonomic mapping and feature in it, but not neatly, often appearing in more than one place. That is, my general taxonomy certainly does not reduce to categories informed by these theoretical orientations,

> Technocratic rationality
> Reactionary modernism
> Defence of humanism/anti or post-humanism
> Primitivism
> Defence of tradition
> Theological objections
> Unintended consequences
> Computers as political technology
> Moral objections

Figure 2 A taxonomy of theoretical positions

indeed it *complicates* them. All these other groupings, lists joined by what Bogost (2012) called the 'gentle knot' of the comma, can be unjoined and redistributed across all of the eight categories of my overarching taxonomy, where they will not sit still and do not fit neatly. They loosen it up, let it breathe, perhaps tend to let it begin to transform; the taxonomy after all is both a catalogue and a list, and the latter demands endless and ongoing elaboration. As Georges Perec, a theorist of everyday life and a lover of the automated sort, noted:

> [N]othing seems simpler than making a list, but in fact it's much more complicated than it seems: you always leave something out, you're tempted to write etc., but the whole point of an inventory is not to write etc. (Perec, 2009)

Critical orientations and situations

I now briefly turn to reflect on my own situated position – and my own take on the computational. My intention has been to develop a cultural study of and with anti-computing that is orientated by broadly historical materialist approaches and that substantially engages with media archaeology and medium theory. This general outlook produces certain starting points. Notably it generates an account framed in material and materialist terms, cognisant of overarching social and techno-social logics, understood or explored in their diachronic as well as synchronic aspects; the complex temporalities and continuities and disjunctures of the computational/anti-computational demand this. Technology, recognized in its materiality, is always also explored in relation to culture, and in relation to social power; that is to say, by a critical theorization of technology and culture that sees the two as co-evolving. This general theoretical/critical approach does not demand or entail *either* a sustained rejection of the technological, or its vehement adoption. Finding a reconfigured role for technology in the generation of new forms of future possibility – which might stand against the expectations of progress – is certainly not a matter of quantity, more of this stuff or less or that. However, I recognize that where I write from does have consequences. I have more sympathy with some of the anti-computing formations I explore than with others, and perhaps also read their significance differently.

This account turns around shards and fragments of computational culture, involving therefore a series of decisions about what to pick up and what to leave on the ground. For instance, thinking about questions of technologies and bodies, I selected as one object of inquiry the science fiction (SF) writing of Hanu Rajaniemi, read as a defence of embodiment over anti-human uploading fantasies, choosing to explore this rather than – say – concentrating on a discussion of the moral economy of the games-censorship lobby in the UK in the 1990s; Either of these could find a place in the kind of expanded taxonomy of anti-computing under development here. To me, the first seems to provide a more intriguing way of thinking through unease around 'artificial' bodies and practices than the second. Similarly, whilst much public attention is being paid to issues falling within the category I designate 'horrible humans' – cyber-terror, internet pornography, shaming are often accounted for currently in terms of the unleashing of the inhuman in the human, of letting beasts not gods into the machine – this has not been explicitly addressed through a case study; rather, I return to it in the concluding chapter to ask why and how it has exploded as an operative category and why, as a categorizing operation, it has become so pervasive and so effective in ascribing blame. Other shards were more demanding of attention.

Am I anti-computing? I have elsewhere considered the possibilities of silence and refusal and have certainly critiqued particular computational formations in terms of their neoliberal instantiation, amongst other things (Bassett, 2007, 2013, Bassett and Roberts,

2020). However, against the writing on the computational framed in terms of refusal that I have produced, I would set other work that articulates a conviction that there are other forms of the computational than those that dominate today, and that these are important, significant, creative, affirmative, live giving, and can be used as tools for justice. So, I am not declaring for anti-computing. And this is not a manifesto. Nor is it a call for de-acceleration, for a go-slow, and certainly it is not a call for revived forms of primitivism. On the other hand, an anti-computing impulse of a kind chimes with my desire to upset an established form of thinking about technological innovation, precisely the teleological progress narrative already outlined above, and to protest the multiple ways it is put to ideological use. Anti-computing opens a way to think critically about the claims and instantiations of the computational, and as part of this I seek to think about the discrete formations I explore in critical terms. So, if this is not a manifesto, it is, as Barad would put it, a cut (Barad, 2003). And I am intending to cut in.

To attempt a summation of the issues opened up here, which stand as a rationale for what is to come, and which also set out my own position, I close this chapter by suggesting that anti-computing is useful to develop and explore for six connected reasons.

- Hostility to the computational is a significant strand in the fabric of human–computer engagement as that tissue has developed over seventy years. It is a strand that is largely ignored in histories, and certainly rarely viewed as something systematic. It is worthy of attention because of this. It offers a different perspective on, and suggests other ways to gauge the implications and claims of, computational 'advancement'. Anti-computing might be, somewhat in the manner of counter-factual history, deployed to raise the possibility of other possibilities.
- Anti-computing moments are part of the history of computing as a material cultural history and are worth excavating, in and of themselves.
- Anti-computing formations are at times strident and gain much popular acceptance. But they are also, measured against the overwhelming tide of acceptance of the computational insertion into everyday life, relatively uncommon …

- ... But although there has never been a grand movement, an overarching narrative of computational dissent, nor a single defining moment that has persisted, various arguments recur, repeating and also mutating in each iteration. These circuits, giving insight into processes and forms of intersection between material cultures and their histories and social and political developments over time, demand further investigation.
- Exploring anti-computing can contribute to developing ways of critically understanding media-technological histories – which I read as asking questions about the intersection of questions concerning computational technologies, culture, and power.
- Anti-computing formations do not necessarily set out to directly challenge the power structures within which they are embedded; their orientations are varied. But exploring anti-computing formations can expose the ideological power and force of authorized versions of computational 'advancement' that constitute the dominant computational imaginary. The anti-computational turn which I make in this book therefore has critical intent.

Notes

1 The Luddites' rational, if impassioned, arguments against the use of machines as political tools by another class, expressed as machine breaking (and had not that other class also 'spoken' through those same machines?), were supposed to be senseless.
2 Some confirmation of this comes via Google's Ngram, which shows the term peaking in the year 2000 (time span 1980–2015).

Bibliography

Anderson, Chris. 2008. 'The End of Theory: The Data Deluge Makes the Scientific Method Obsolete', *Wired*, 16 July, http://archive.wired.com/science/discoveries/magazine/16–07/pb_theory.

Andrejevic, Mark. 2015. 'The Droning of Experience', *The Fibreculture Journal* 187, http://twentyfive.fibreculturejournal.org/fcj-187-the-droning-of-experience/

Arendt, Hannah. 1998 [1958]. *The Human Condition*. Chicago: Chicago University Press.

Barad, Karen. 2003. 'Posthumanist Performativity: Toward an Understanding of How Matter Comes to Matter', *Signs: Journal of Women in Culture and Society*, 28(3): 801–831.

Bassett, Caroline. 2007. *The Arc and the Machine: Narrative and New Media*. Manchester: Manchester University Press.

Bassett, Caroline. 2013. 'Silence, Delirium Lies', in Geert Lovink and Miriam Rasch (eds) *Unlike Us Reader: Social Media Monopolies and Their Alternatives*. Amsterdam: Institute of Network Cultures, 146–159.

Bassett, Caroline. 2017. 'What Perec Was Looking for: Notes on Automation, the Everyday, and Ethical Writing', in Rowan Wilken and Justin Clemens (eds) *The Afterlives of George Perec*. Edinburgh: Edinburgh University Press, 120–135.

Bassett, Caroline and Ben Roberts. 2020. 'Automation Now and Then: Automation Fevers, Anxieties and Utopias', *New Formations*, 98: 9–28.

Bassett, Caroline, Ed Steinmueller, and George Voss. 2013. *Better Made Up, Science Fiction and Innovation*. Working paper to NESTA.

Bauman, Zygmont. 2003. *Wasted Lives: Modernity and Its Outcasts*. London: Wiley.

Beck, Ulrich. 1992. *Risk Society: Towards a New Modernity*. London: Sage.

Berry, David M. 2011. 'The Computational Turn: Thinking about the Digital Humanities', *Culture Machine*, 12: 1–22.

Bey, Hakim. 1991. TAZ: *The Temporary Autonomous Zone*. London: Autonomedia.

Blas, Zac. (2012–14) 'Facial Weaponization Suite', www.zachblas.info/works/facial-weaponization-suite/.

Bogost, Ian. 2012. *Alien Phenomenology, or What It's Like to Be a Thing*. Minnesota: University of Minnesota Press.

Borges, Jorges. 2000. *Labyrinths*. London: Penguin.

Cockburn, Cynthia. 1992. 'The Circuit of Technology: Gender, Identity and Power', in Eric Hirsch and Roger Silverstone (eds) *Consuming Technologies, Media and Information in Domestic Spaces*. London: Routledge, 32–48.

Cramer, Florian. 2014. 'What is "Post-digital"?' *APRJA Post-Digital Research*, 3(1).

Doane, Mary Anne. 1990. 'Information, Crisis, and Catastrophe', in Patricia Mellancamp (ed.) *Logics of Television: Essays in Cultural Criticism*. Indianapolis: Indiana University Press, 221–238.

Erkki, Huhtamo. 1997. 'From Kaleidoscomaniac to Cybernerd: Notes toward an Archaeology of the Media', *Leonardo*, 30 (3): 221–224.

Evans, Brad. 2019. 'Histories of Violence: Thinking Art in a Decolonial Way', interview with Lewis R. Gordon, *Los Angeles Review of Books* (June).

Fazi, M. Beatrice. 2018. *Contingent Computation: Abstraction, Experience, and Indeterminacy in Computational Aesthetics*. London: Rowman & Littlefield International.

Foucault, Michel. 1972. *The Archaeology of Knowledge*. London: Routledge.

Foucault, Michel. 1997. 'What Is Critique?', in Sylvère Lotringer and Lysa Hochroth (eds) *The Politics of Truth*. New York: Semiotext(e.).
Gebru, Timnit. 2021. On the Dangers of Stochastic Parrots: Can Language Models Be Too Big? *ACM Conferences*, FAccT'21, https://dl.acm.org/doi/10.1145/3442188.3445922.
Haraway, Donna J. 2016. *Staying with Trouble: Making Kin in the Chthulucene*. Durham, NC: Duke University Press.
Harrison, Emma. 2020. 'Activism, Refusal, Expertise: Responses to Digital Ubiquity'. Doctoral thesis, University of Sussex.
Hartmann, Maren. 2009. 'After the Mobile Phone?' in M. Hartmann, P. Rössler and J.R. Höflich (eds) *Social Changes and the Development of Mobile Communication*. Berlin: Frank & Timme, 35–54.
Hawking, Stephen et al. 2015. *Research Priorities for Robust and Beneficial Artificial Intelligence*. Open letter, https://futureoflife.org/ai-open-letter).
Hayles, N. Katherine. 2008. *Electronic Literature: New Horizons for the Literary*. Notre Dame: University of Notre Dame Press.
Jameson, Frederic. 2015. 'The Aesthetics of Singularity', *New Left Review*, 92: 101–132.
Jordan, Tim. 2008. *Hacking: Digital Media and Technological Determinism*. London: Polity.
Jordan, Tim. 2017. 'A Genealogy of Hacking', *Convergence*, 23(5): 528–544.
Joy, Bill. 2000. 'Why the Future Doesn't Need Us', *Wired*, 4 January, www.wired.com/2000/04/joy-2/.
Lachney, Michael and Taylor Dotson. 2018. 'Epistemological Luddism: Reinvigorating Concept for Action in 21st Century Sociotechnical Struggles', *Social Epistemology*, 32(4): 228–240.
Lanier, Jaron. 2010. *You Are Not a Gadget*. New York: Knopf.
Latour, Bruno. 1991. 'Technology is Society Made Durable', in John Law (ed.) *A Sociology of Monsters: Essays on power, Technology And Domination*. London: Routledge, 103–131.
Latour, Bruno. 2011. 'Love Your Monsters: Why We Must Care for Our Technologies as We Do Our Children', *Breakthrough Journal*, 2: 19–26.
Morozov, Evgeny. 2012. *The Net Delusion: How Not to Liberate the World*. London: Penguin.
Pariser, Eli. 2012. *The Filter Bubble: What the Internet Is Hiding from You*. London: Penguin.
Perec, Georges. 2009. 'Notes on the Objects to be Found on My Desk', in *Thoughts of Sorts*. London: David R. Godine.
Silverstone, Roger. 1994. *Television and Everyday Life*. London: Routledge.
Smith, James H. 2011. 'Tantalus in the Digital Age: Coltan Ore, Temporal Dispossession, and "Movement" in the Eastern Democratic Republic of the Congo', *American Ethnologist*, 38(1): 7–35.
Solnit, Rebecca. 2007. *Storming the Gates of Paradise*, Cambridge: Cambridge University Press.
Solnit, Rebecca. 2013. 'Google Invades', *London Review of Books*, 35(3), 17 March.

Steyerl, Hito. 2013. 'Too Much World: Is the Internet Dead?', *e-flux*, 49.
Stiegler, Bernard. 2013. *What Makes Life Worth Living: On Pharmacology*, Oxford: Polity.
Traub, Valerie. 2007. 'The Present Future of Lesbian Historiography', in George E. Haggerty and Molly McGarry (eds) *The Companion to Lesbian, Gay, Bisexual, Transgender, and Queer Studies*. Oxford: Blackwell, 124–145.
Turing, A.M. 1950. 'Computing Machinery and Intelligence', *Mind* (October): 433–460.
Virlio, Paul. 1997. *Open Sky*. London: Verso.
Webster, Frank and Kevin Robins. 1986. *Information Technology: A Luddite Analysis*. London: Ablex Publishing Corporation.
Weizenbaum, Joseph. 1976. *Computer Power and Human Reason, from Judgment to Calculation*. San Francisco: W.H. Freeman.
Wiener, Norbert. 1950. *The Human Use of Human Beings*. London: The Riverside Press.
Wiener, Norbert. 1961. *Cybernetics, or Control and Communication in the Animal and the Machine*. Cambridge, MA: MIT Press.
Wing, Jeanette M. 2006. 'Computational Thinking', *Communications of the ACM*, 49(3): 33–35.
Winner, Langdon. 1985. 'Do Artifacts Have Politics?', in D. MacKenzie and J. Wajcman (eds) *The Social Shaping of Technology*. Milton Keynes: Open University Press, 26–38.
Yusoff, Kathryn. 2018. *A Billion Black Anthropocenes or None*. Minneapolis, MN: University of Minnesota Press.
Zielinksi, Siegfried. 2008. *Deep Time of the Media*. London: MIT Press.
Zylinksa, Joanne. 2009. *Bioethics in the Age of New Media*. Cambridge, MA and London: MIT Press.

Chapter 2

Discontinuous continuity: how anti-computing time-travels

> The use of concepts of discontinuity, rupture, threshold, limit, series, and transformation present all historical analysis not only with questions of procedure, but with theoretical problems. It is these problems that will be studied here. (Foucault, 1972: 21)

What began as a cultural and medium-theoretic study roving across a series of sites increasingly involved engagement with issues of computer history; this came to seem essential to understanding and constructing anti-computing. But why look back? Following the turn of the decade, as I complete this book, there is abundant hostility to computing, anxiety about its impacts, and rejection of its visions in the here and now. Automation anxieties around the future of work are fuelling a new anti-computational turn, data-surveillance issues haunt formal politics, and there is rising concern over screen 'addiction' in the young. Limit points are being declared and last-chance saloons announced. The massacre in Christchurch, New Zealand in 2019, streamed live for a cruelly long time and endlessly disseminated, seemed to many to at once exemplify everything wrong with the platforms – deemed uncaring, unable to control what they unleashed, and not choosing to do so since their end goal is profit, not social well-being; and to point to everything wrong with digital humans, whose capacity to share ugliness and to share extremism appears to expand in tandem with the expansion of the means to do so. Since then, not much has changed, although that event has receded from consciousness – unconscionably quickly, perhaps. So, why not stay in the here and now and explore contemporary hostility? My response is that to remain entirely within the present and/or within near future horizons (real and imaginary) risks succumbing to forms of presentism prevalent both within the computational mainstream and within many of the anti-computational formations investigated here. These forms of presentism are questioned in this book.

So, as well as being a cultural study, this book is a cultural history, albeit of a discontinuous kind. The hope is that historical inquiry re-energizes explorations of the contemporary condition because it offers a route through which to examine the 'plastic' (Uprichard, 2013) forms of presentism that many have identified as part of the current 'conjuncture' or moment (e.g. Liu, 2004, Jameson, 2015). Presentism, explored further below, is a key element of computational capitalism. It has consequences for theorizing and understanding anti-computing. An exclusive focus on the present renders hostile responses to the computational into discrete and proximate issues (problems with this kind of corporation or this kind of architecture, this new technology or that new behaviour, horrified responses to this kind of screening, or that kind of digitally mediated hatred) cutting them off from larger, overarching formations and short-circuiting longer, more complex histories of refusal, critique, unease, anxiety, and activity.

Consider three sets of more or less hostile responses to the computational that have circulated in the decade since 2010, grouped together here because each declares that what it excoriates is *novel*. The first of these groupings is anti-platform, and increasingly anti-monopoly; it finds a key role for social media platforms in the crisis of democracy represented by Trump, and Brexit and its aftermath, locating this in the capacity of platforms to host and promote extremism and to enable fake news (the New Zealand case exhibits these logics). The second form of dissent, markedly critical of how computers are being used by nation-states, and of what they are being used to do to ordinary people, circulates in the post-Snowden era and is concerned with personal data in an era of 'Big Data' – the latter a label that itself makes claims about novelty. A third response is the postdigital analysis (Cramer, 2014). This assesses the state of the digital itself and is in this sense a meta-analysis. It is above all jaded and 'disenchanted' (Couldry, 2014, Bassett, 2015). It says that the digital, now instantiated as a pervasive material of the everyday, has lost its purchase as a defining cultural logic; it continues, and yet it is no longer really new or interesting, nor does it open new horizons. This has sometimes produced an aesthetics challenging invisibility and/or the normalization of pervasive mediation – Hito Steyerl's (2013) work and writing is germane here, but more often has fuelled a turn to a *new* new attraction. In the case of the postdigital there is still a new claim – since what is said to *be* new is the claim to go beyond the prioritizing of the new. As an aesthetic, the postdigital is thus hostile to 'the digital'.

These three examples of anti-computational formations operating more or less in the present are easy to classify as newly arising, and are often seen simply as new. But all three also relate to earlier irruptions, from which they are now largely *delinked*. This is not exceptional; there are many forms of anti-computing that are not entirely new, and that may even be recognizable as familiar, even whilst this 'familial' relation is disavowed, set aside, or forgotten. Examples might include that falling away of enthusiasm for the fully computerized society that led Edmund Berkeley, editor of *Computers and Automation*, one of the first computer magazines,[1] to change the title to *Computers and People*. This is an early example of computational disenchantment based on dehumanization concerns, but it is not a formation that tends to be linked with current concerns around dehumanization and automation laid at the feet of contemporary AI. Or consider the 2001/2 bursting of the dot.com bubble, which deflated a form of net euphoria partly by identifying it as an (unjustified) euphoria about the new; the postdigital formation, clearly linking in to this earlier moment, isn't entirely novel, even if it often presents itself as such.

Then there are the anti-computing sentiments threading into the counter-cultural movements of the 1960s and centrally concerned with the rise of data and the perfection of the state 'machine' (Turner, 2006). The issues raised around that time have not been resolved, although neither have the most apocalyptic visions of a fully computerized world arisen (yet). The critique faded away, or, as the other kind of weathermen say of a storm, it lost its *identity*. Or consider the widespread opposition to the Regulation of Investigatory Powers Act in the UK in the early 2000s (John Naughton's commentaries in the *Observer* provide an indicative flavour of this). The investigatory powers legislation, which was passed into law, plays its part in contemporary surveillance regimes. The UK has since completed the implementation of what many regard as an extension to its terms,[2] rather than a simple renewal of some of the powers it legislated for. This has not gone forwards without hostile commentary, but even within that, the widespread opposition, unease, and hostility provoked the last time around, substantial in liberal, radical and mainstream media and amongst campaign groups, has far more rarely surfaced in public debate. Finally, consider the many late 20th-century campaigns against camera surveillance in public spaces in the UK and elsewhere. These were both local and national, and evidenced a widespread unease. When did that switch in attitudes which made it possible to market general surveillance as a necessary, if not desirable, part of public space come about?

The mismatches of scale and register in these examples are intentional. What links them is that they *partake in the same logics of recurring appearance and disappearance*. What might be termed, adapting Foucault (1972: 209), the *dissolvability* of anti-computing formations, their propensity to subside and vanish from view, even whilst remaining in the mix, is marked.

More examples: who would not find the notion of demanding an iron apron to word-process strange today? Iron aprons were a union-based response to early VDUs (visual display units) and to concerns with the health hazards screens might pose to pregnant women, and although more recent concerns with blue light recapitulate earlier fears about VDUs and eye strain, screen technology changed and the pregnancy concern died away. This indicates how older dissent may be delinked from contemporary issues, deemed irrelevant because the specific technologies that provoked them are 'out of date'. A different issue is whether the critique of social relations which they entailed becomes irrelevant in relation to successor technologies and the dissent they afford. Here are two examples where the impetus for earlier hostility fading is less certain: first, what happened to opposition to Photoshop and digital photography from those in favour of 'real' photography as the guarantor of indexical realism and documentary 'truth'? This was a debate held in professional photography circles, amongst other places, in the 1990s. Clearly, the stakes of this earlier discussion connect to unease around deep fakes today; this despite the fact that material indexicality is no longer the issue around which debate circulates.[3] Second, automated passport gates. Were you against them once? Did you view your objections as principled at that time? Many did. Do you remember why, or when, you changed your mind, or set aside your objections? Moreover, assuming your original objections *were* set aside, were they set aside for good? Or do they remain somehow with you, might they be refound as a set of responses that might inform a response to different modes of surveillance now arising? How does personal acceptance relate to a general hardening of borders, to populist nationalisms?

These examples indicate the fragility, and even the ephemerality (Chun, 2008a), of many forms of computational dissent, which are of their moment, and, to some extent, live in it (which is to say they also partake in a form of presentism). But they also evidence a paradoxical durability, obduracy, and persistence. Even when apparently terminally discarded because attached to an outmoded technology, or because attacking an outmoded problem, many anti-computing formations remain if not immediately ready at hand, then apparently capable of being reached for, invoked, ready to *inform* a new landscape. What is dissolved *in*, is not necessarily dissolved *out*. One reason, then, to explore anti-computing formations across time is to find a way to avoid being consumed by the *force* of presentism; but another is to register the limits of its operations, the limits of what it may suppress of the past. Walter Benjamin's sense of history as unfinished resonates here, with its argument that what appears to have been terminally consigned to a dead past, so that its place in history, and its significance, have been settled (and, in general, by the victors), may be unsettled, and may become active in the present (Benjamin, 2006).

This suggests to me two priorities. The first is to refind anti-computational formations in the past, and the second is to understand the dynamics of their recurrence. I want to draw out and understand a series of anti-computing formations, and to understand them both as discrete and as connected, as formations that, more than is usually understood, persist across time, and that might rise, fall, and revive, in non-linear, complex, and indirect ways.

This contrasts with linear accounts of the computational, which view it has having risen, apparently inexorably, to become a key operational logic in the cultures of late 20th- and 21st-century global capitalism; to have become that which organizes the present and that which constitutes the 'superhighway' to the future (the old 1990s term is apt to describe an ideological orientation, even if it never fitted the reality). This claim (highly ideological) is compromised once the anti-computational is taken into that account. The trajectories and circuits of the anti-computational are at times hardly to seen, at others brilliantly lit. They often appear disconnected, or produce nodes that seem to have risen autonomously, from nowhere at all, or as the direct consequence of something 'all new' (AI, automation, platforms, for instance), but they also seem peculiarly familiar, to have recourse to a series of recurrent and pre-existing tropes, or to make connections with earlier sensibilities or formulations. *The central problematic of this chapter is how to identify and account for the forms of disjunctive continuity that characterize anti-computing as a material cultural form and that give it its 'history'.*

Doing 'computer history'

The engagement between information technology and culture central to mid- to late 20th-century and early 21st-century life is complex, operating at a dizzying series of scales and registers, involving heterogeneous actors – humans, technologies, techniques, materials, environments – bound into multiple overlapping networks, constituted and reconstituted over time, in relations of radical and structural asymmetry. Computer historian Paul Edwards rightly observes that writing an entire history of computing would be 'a colossal and difficult task, beyond the reach of any individual' (Edwards, 2001: 87). The question, then, is how to make the cut.

Big stories of computing tend to be prey to a series of occlusions, often arising through forms of substitution.[4] Complex formations are replaced by abstract technologies, or 'wizard' inventors producing characteristic distortions across a range of technical histories (Edwards, 2001: 88). Such distortions arise when key individuals are invoked, not to 'stand for' or symbolize a larger formation as a particularizing synecdoche (Whitsitt, 2013), but as having *themselves determined* the history of a corporation; Steve Jobs has figured in this way, for instance. 'Great men' may also be replaced by 'great software' (Photoshop, for instance), or great algorithms (Google springs to mind). Institutional histories can produce less myopic accounts; a successful example is Steven Levy's history of Apple's early decades (Levy, 1994), but when institutional psychologism replaces the individual psychologism that informs wizard accounts and that is translated into a form of object agency, other problems arise. An insistent focus on how a company 'thinks' or performs its identity easily comes at the expense of what it *does* (its technologies, its relations of production), and even of its real location. Relevant here are accounts that neglect the working conditions of the industrial plants of the global South and the repatriated low-wage service centres inside the US in favour of poring over the temple architectures of the giant corporations in Silicon Valley in search of the 'DNA' of their owners and the character of what is (only very partially) made within their grounds. To read a homogeneous 'global' narrative of computerization off from this central dashboard is to misrecognize the significance of the location of material (as well as immaterial) production, the specific experience and use conditions of billions of instantiated users, and the complexity of the flow of technologies, ideas, and workers to and from the dominant hubs. Indeed, it renders computer culture 'Californian', perhaps leaving a small space for 'European' thought as its mordantly critical other, as Richard Barbrook notoriously suggested in the 1990s. Barbrook's analysis, radical in intent, is now striking for what it excludes. Where is the rest of the world here? Another consequence of this kind of ordering is a tendency to confine origin stories of *resistant* practices (e.g. hacking) to the same core region, when, as Kat Braybrooke notes, they have always been more widespread (Braybrooke and Jordan, 2017). The same tendency can be observed in (congruent) histories of cyber-feminism, always more than Anglo-American, Australian or European, but all too often compressed into that mould.

Making the cut: or why social history versus medium specifics isn't good enough

Lisa Gitelman's rejection of heroic stories of 'how one technology leads to another, or of isolated geniuses working their magic on the world' (Gitelman, 2006: 7), translates some of these issues into a narrower register – that of media/medium studies. Gitelman argues that technological media demand social and cultural histories. Making this case she is essentially mapping an established 'media history' approach, critical of populist hero histories but also of what it discerns as forms of reductive technological determinism, onto 'new' digital media history. Gitelman's work is representative of a form of media history that critiques technological abstraction whilst demonstrating an awareness of medium operations. Another example is found in Kate Lacey's (2013) work on early radio, which fuses sound studies with institutional histories. Both scholars help to make the case for demanding (more) recognition of the social and cultural conditions of possibility within which digital technologies, and the political economies they foster, come into being. This general approach has produced important histories of various aspects of the computational – for instance, of the role of IBM in the Second World War, of cybernetics as Cold War technology, of the technologies of everyday life in the constitution of American society/domestic life in the 1950s, of broadcasting institutions such as the BBC and others (see Gerovitch, 2001, Hendy, 2000, Black, 2011).

None of the writing invoked here is entirely social constructivist; indeed, most of it might be characterized, following Skågeby and Rahm's (2018) useful definition, as post-constructivist. Partly because of this, perhaps, it is rarely polemical about the virtues of its own method. It is nonetheless taken up as a target and used by new materialist critics of 'traditional' media-historical approaches, who (by framing it as such), argue that if attention *is* paid to medium matters, not *enough* attention is paid. This is felt to produce occlusions. Gitelman's rejection of teleological stories is based on her sense that 'media are unique and complicated historical subjects' (Gitelman, 2006: 7), and this complexity is recognized to inhere in part in their materiality. But for new materialist critics of 'traditional' ways of doing history this admission of the material may not suffice. What needs challenging, from their perspective, is that tradition which holds humans to be the history makers in the final analysis. For various flavours of German medium theory, notably, the conditions of possibility for agency and the nature of subjectivity itself are reconstituted by technical media.

Cuts and lines

This suggests a blunt division in the orientation of theoretical work, sometimes framed as a division between a focus on materiality or on representation, which needs to be questioned. This might begin by acknowledging the distances between, for instance, Leslie Haddon's engagement with domestic histories and gaming in the 20th-century UK (Haddon, 1988), Fred Turner's work on techno-cultures and Silicon Valley (Turner, 2006), and Kittler's essay on 'Protected Mode', a classic of medium history (Kittler, 1997), which are substantial. On the other hand, proclaimed absolute divisions between various forms of digital media research made by proponents of one side or the other do not necessarily reflect the opposing view in its complexity, nor indeed acknowledge the ambiguity and contradictions of their own positions, and may usefully be challenged.

Anti-computing is part of that challenge. If histories are cuts (and must be cuts, given the impossibility of completeness), then anti-computing is a cut. It is a *cutter* too, since with it I want to make particular kinds of incisions, to divide and to bring together. But cuts do not have to produce binary polarities. Cuts join, and there is always what happens along the line of the cut itself, and in the liminal zones around it, where contagion happens. Perhaps, as Ann Light suggests in her work on technology and subjectivity, there are ways of cheating the cut (Light, 2011). Light is drawing on China Miéville's novel *The City and the City*, in which rival polities overlap but also disavow one another's existence, producing an impossible but also real separation. The protagonist finds a way out of both cities by exploiting or traversing the cut that joins them. 'I was learning … how to walk between, first in one, then the other, or in either … [a] covert equivocation' (Miéville, 2009: 368). 'To walk between' different conceptions of media history is to challenge absolute divisions, all too often erected, between the social and the technical, and between 'media history', 'medium studies', and media archaeology. The point is to produce an account that is *not* reducible to technologies presumed to 'determine' our situation (Kittler, 1997), *nor* to institutional readings, whilst *also* disturbing accounts cleaving to representation rather than material that flatten the technological or render it into discourse. Anti-computing as a cut, a walk-through that gathers what it needs, is to be organized by a reading of the computational as a process of co-evolution between machines and humans and therefore as intrinsically (in its materialized and operating form) techno-social. In what follows this line is walked.

Archaeology and media archaeology

In the *Archaeology of Knowledge* Foucault (1972) explores distinctions between two forms of history. The first is a seeking for origins, an anthropological history, that which seeks in history to find reassurance of the sovereign subject, which works through this subject and presumes their coherence across time, an approach generating origins and positing a form of linearity which also indicates teleology. The second form of history is based on rupture and discontinuity. Championing the second form (and binding this to an anti-human reading of Marx, to Nietzschean genealogy, and to psychoanalysis, all of which are viewed as engaging in the decentring of the human), Foucault avowedly sets out not to attack history, but only a certain *form* of history. Moreover, he claims that his intervention is part of a larger shift, an 'autochthonous' transformation in the thinking of many scholars or 'the eclipse of that form of history ... secretly ... related to the synthetic activity of the subject' (Foucault, 1972: 14). Other disciplines, those which 'evade the work of the historian', are invoked, those which are concerned with rupture, and which search for 'displacements' and 'transformations' of concepts. Foucault notes emerging forms of the history of science, particularly those influenced by Gaston Bachelard, lauding such endeavours as a 'new form of history' at whose heart is 'the questioning of the document' (Foucault, 1972: 6).

Foucault writes that in these conditions (in this time of transformation) discontinuity demands reassessment. It is no longer a problem, a gap, something to be plugged in(to) a linear account. On the contrary, discontinuity should be focused upon by the historian, who may now explore 'the limits of process, the point of inflection of a curve, the inversion of a regulatory movement, the boundaries of an oscillation'. Discontinuity becomes 'an instrument and object of research' (Foucault, 1972: 9). The distinction is between a history that provides for the subject (and renders history itself as) 'a place of rest, certainty, reconciliation, a place of tranquilized sleep' (Foucault, 1972: 14) and that new form of history, founded in rupture and focusing on discontinuity rather than linearity, that Foucault wished to develop. The flat line of the long sleep is to be interrupted. Something is to be jolted into new life – but vitality will not come from the old sovereign actors. It was the provision of a guaranteed place for them that produced history *as* sleep.

Object and instrument; discontinuity is instrumental because it is the blade that may shuck out the matter or material documentation that, says Foucault, must be grasped, explored, and worked upon. The goal is no longer to reconstitute 'on the basis of what the documents say, and sometimes merely hint at ... the past from which they emanate, and which has now disappeared far behind them'. Documents are no longer to be understood as 'inert material' which can only 'refresh' memory. History is 'the work expended on material documentation', and the point is 'to work on it from within and to develop it'. History also now defines 'within the documentary material itself unities, totalities, series, relations'. It is in this way that we can make sense of Foucault's declaration that documents become monuments ('documents into monuments'), so that 'in our time history aspires to the condition of archaeology, to the intrinsic description of the monument' (Foucault, 1972: 7–14).

The *Archaeology of Knowledge* is concerned not only with disjuncture defined as pure schism, but also with generating new connections. Refusing the latter entirely would, it is acknowledged, leave *only* 'a plurality of histories'; the distinction is between a total history and a general one. The blade here is subtle, it cuts, but, in cutting, it also sutures. Foucault is interested in dissecting 'systems of relation' between series, in the conditions of possibility that dictate 'what series of series' may be established, and in 'what form of relation may be legitimately described between ... different series' (Foucault, 1972: 10). Thinking in tabulated form he articulates this as a question about what tables it becomes 'possible to draw up'.

If the arguments of the *Archaeology* do not espouse absolute relativism or disjuncture, there is also an explicit refusal to frame the archaeological approach as simply a structuralist mode of doing history (or philosophy, or the history of ideas, for that matter). The question of legitimation (legitimate description), and the kinds of investigations this approach might suggest (including taxonomic work), needs to be understood, Foucault argues, in far more than purely formal terms. The *Archaeology* remains a critical project and resonates with his other writings on governmentality. The king, as he declared in *Power/Knowledge* must still have his head chopped off if we are to change how we are governed (Foucault, 1980).

Media archaeology: priority and determination

The archaeological intervention has latterly been felt with some force in media and cultural studies, in software or code studies, and via a somewhat discrete mode of media archaeology developed in cinema studies by Thomas Elsaesser (2004, 2016). One core activity has been in medium theory where Jussi Parikka has done much to define its activities and orientations (see later). Media archaeology has gathered under its aegis a wide set of writings and writers, some adopted retrospectively as progenitors – McLuhan, Mumford, and Ellul are often invoked, perhaps in ways that would make them uneasy, had they any choice in the matter. Media archaeology tends to be defined as a collection of related approaches rather than a single discipline (Parikka, 2012), but as a whole the project is deeply in debt to Foucault's archaeology, and influential practitioners of *digital* or *computational* media archaeology have found inspiration in Foucault's insistence on discontinuity, and his demand that attention be paid to series and strata, and their relations, rather than deferring to a presumed external ordering or system imposed from the outside. A reviewer of Parikka's work summed up a commonly held view by declaring media archaeology a 'successor variant' of Foucault's archaeological project (Anthony, 2012). Some practitioners, however, set aside the Foucauldian connection, with its insistence on power and domination, preferring the temporal politics of Walter Benjamin, or focusing on forms of direct experimentation and remaking (see Goodall and Roberts, 2019).

Media archaeology is a response to the challenge of writing media histories, a search for an approach and/or method that can adequately take account of the circular, the recursive, the submarining, the looped – these dynamics and characteristics being read *as* the temporalities of the machine, or as intrinsic characteristics of technical media. A fault in 'standard' media histories is summed up as their tendency to *resolve* the complex temporalities and simultaneities of technological operations into standard linear time (lines). Doing so, it is argued, means both that technical operations are not accounted

46 *Anti-computing: dissent and the machine*

for and that (therefore) the consequences of the instantiation of these technologies as cultural techniques, or as pervasive cultural logics, cannot be understood. Thus Geert Lovink characterizes the media archaeologist Wolfgang Ernst's work as a mounting a direct challenge to 'the usual chronological reading of media, from photo and radio to television and the Internet' (Lovink, 2003).

The critique is above all topical. The rise of technical media means rewriting history and/or demands finding new ways in which it may be undertaken. For media archaeology advocates, older theoretical frameworks are strikingly out of place in the contemporary world. Standard media studies are charged with an obsessive attachment to linear development, said to be a consequence of the desire to map a human narrative of progress at the expense of considering what other forms of ordering are out there – orderings which might come *prior* to what is viewed as a 'reduction' into narrative and its temporalities. This approach is said to strip out the complexities of technological orderings and temporalities. In response, the media archaeological project seeks to understand what other possible patterns emerge if the lens is widened, the perspective altered, and narrative organization set aside.

The above is in danger of over-compressing the multiple approaches of media archaeology. This media-archaeological thinking that focuses on priority, linearity and recurrence, and significance is now looked at in more detail, being found to offer a resource to assist in the project of doing anti-computational history whilst also being found to be problematic. This is a friendly critique, however. It is undertaken, as Skågeby and Rahm (2018) put it in their account of feminist media archaeology, with the intention of being useful.

Ernst, priority (and the material conditions of legibility)

Friedrich Kittler, the German medium theorist often regarded as a father to media archaeology (an aptly patriarchal label, given his gender politics), exploring cinema, gramophones, and typewriters, famously argued that the media have come to determine the situation (Kittler, 1997). Elsewhere, he took on questions concerning computing more directly, arguing that the dilemma between 'code and language … seems insoluble', so the program will run when the programmer's head is empty of words (Kittler, 2008: 46). The corollary of this might be that language may come (or be heard) only after code stops running. Echoes of Kittler can be heard in the account of computational temporality developed by Wolfgang Ernst, who makes strong claims that in the time of technical media new forms of inquiry are needed. Ernst's work is avowedly (also) in the lineage of Foucault (see Parikka, in Ernst, 2013: 4) and Ernst's M.*edium F.oucault* (2000), and he retains a central interest in questions of power. But if Ernst is an heir to Foucault's archaeology, he also breaks with it, replacing the document with the material. For Ernst, media now condition 'the way in which we know things and do them – knowledge and power' (Parikka, in Ernst, 2013: 6). In doing so, he is not alone. As Lovink (2003) notes, 'whereas Foucault looked into social formations, today's media archaeologists are primarily interested in the (hidden) programs of storage media'.

For Ernst, this produces a project centred on an investigation into the *material* (pre)conditions of legibility, material conditions that temper time and space and determine the character of the place of possibility. This is understood as a historical inquiry, but also as an intervention into history itself, one that grapples with (what Ernst sees as) the transformed conditions of its possible operation. Refusing to flatten technology so that it can become just another element of a history that is smoothly linear partly *because* it dissolves the intractable and non-linear operations of heterogeneous materials in order to produce its text, Ernst mounts a direct 'critique of media history in the narrative mode' (Lovink, 2003). Recognizing that '(the) cultural burden of giving sense to data through narrative structures is not easy to overcome', he nonetheless regards this as necessary. Ernst's work is striking for its formal concentration on questions of temporality that emerge outside of narrative structures. It is the techniques and temporalities of the computational, excavated through an examination of hardware, that inform his explorations of the new 'conditions of the sayable and thinkable'. His media archaeology is defined as 'an excavation of evidence of *how techniques direct human or non-human utterances*' (Ernst, in Parikka, 2013).

How might what Ernst offers be responded to? His approach provides a sense of the complexity of the contemporary temporal order and, in challenging a history purged of the impacts of the operations of technical media, opens the way to undermine the conventional ascription of a narrative of linear progress to computational developments and to techno-social forms. However, one of the dangers of media archaeology is that it tends to essentialize.

The desire to fix on the material can, paradoxically, make technology into something *imaginary*, leaving no space to consider or acknowledge the impurity of the operational. The substitution of an abstract material for a materiality thought impure unless purged of its pesky hybridity produces accounts that ignore the *intersection* of the symbolic and the material, the discursive and the linguistic, the code and the hardware; the complexities of the interacting elements that constitute any computational network in operation, that *handle* mediation. Material specifics are lost as a series of disavowed replacements take place. Kate O'Riordan has aptly characterized 'objects' such as these as 'unreal' (O'Riordan, 2018). That this causes occlusions is evident. Returning to Kittler's proclamation that code and language are inimical, so that the dilemma between 'code and language' seems insoluble (Kittler, 2008: 40), we may note that it entirely steps over the question of *how* code – on its own – signifies. Jaron Lanier has explored this problematic in terms of scale (Lanier, 2010), and Wendy Chun in relation to the gap between the program and the program running (Chun, 2008b: 224). Both their interventions make it clear that it is in process that the disjuncture between the abstract system and its located instantiation is made most visible, and where the mutually informing relationship between the symbolic and material becomes inescapable. Refusing an absolute division between the symbolic and the material therefore seems to me crucial. This also means refusing the absolute priority and that absolutely chronological ordering that is seen in Ernst's work, that which divides the technological (as what came *before*) from the social world (as the social formation) as that which always comes *after* – the irony being, of course, that this ordering is itself insistently linear. One of the consequences of Ernst's argument is that it inserts a prior to the prior; what came before technical determination, how can the genesis of the technical be contemplated? Stiegler's development of co-evolution, which includes an account of originary technicity, produces a useful contrast here (Stiegler, 2013). Even Parikka, a media-archaeological loyalist, talking of myths rather than imaginaries, suggests that Ernst is in danger of 'mythologizing the machine as completely outside other temporalities, including the human', in his attempt to offer 'insight into the *a priori* of writing' (Parikka, in Ernst, 2013: 10).

Linearity and recurrence?

A different media-archaeological intervention is made by Erkki Huhtamo, who seeks to grapple with non-linear and discontinuous histories, these characteristics both informing his theoretical orientation and also discerned as operational temporalities in a multimedia age (Huhtamo, 1997, 1999). His identification of patterned recurrence cleaves rather tightly to, and develops, Foucault's sense of the series and its connections/disconnections. It is developed specifically in relation to media technologies and works through a theory of *topoi* that prioritizes discourse rather than material (here is the break with Ernst).[5]

Parallels between recurrence as Huhtamo explores it and the less *media* archaeological, but nonetheless *archaeological* orientation found in historian Valerie Traub's consideration of the deep history of bodies and sexualities (referenced in Chapter 1 and later explored further) can be drawn here. Demonstrating the recurring 'salience' or intelligibility of various bodily tropes (tropes expressed *as* bodily materializations of various forms of sexuality) across long periods of human history, Traub's focus on what is written on or through bodies indicates ways in which inquiries that focus on discourse may deal – albeit in very different ways, and not in ways likely to satisfy Ernst – with the latter's demand for an inquiry into legibility and its *material* conditions in technical times. Moreover, as is further explored later, both Huhtamo and Traub are concerned to acknowledge continuities *and* simultaneously to comprehend discontinuity and disjuncture.

Significance: media archaeology and its subject choices

How does media archaeology choose its objects? It has a penchant for neglected histories and forgotten or obscure objects or people. Often it selects its research objects in the interests of producing new patterns or circuits, or discerning new connections across categories that are conventionally divided. Zielinski's deep history of devices for hearing and seeing is an example of this (Zielinski, 2006), Parikka's *Insect Media* (2010) another. Then there are studies focusing on small, abstruse, or unlikely elements of assemblages more often and more 'obviously' explored at different scales or in different registers. Ernst is one such practitioner and is reinvoked later. Finally, experimental media archaeology is focused on remaking and building and reusing old technologies now obsolete (Goodall and Roberts, 2019).

Media archaeology has been criticized for these choices, charged with maintaining a preference for 'what matters less' that is voguish, whimsical, or fractious. Even some of its own adherents fear that bad object choices undermine the larger project of media archaeology, defined as the remaking of medium history (e.g. Hertz, 2010). From outside the tent Scott Anthony (2012) is again representative, summing up one vein of external criticism of media archaeology through this scathing description of its 'values':

> The 'what if' of roads not taken is prized over a present whose virtues are assumed to be overstated. To pay dues to the mainstream, to accept at face value, or take common parlance seriously is nearly always to be beneath contempt.

This criticism provokes a defence of media archaeology's choices – and perhaps also of my rationale in working through 'non-obvious' objects in my exploration of the anti-computational. Playing at 'what if' as an end in itself *is* found in media archaeology; however, it doesn't fit as a description of the whole (it might characterize a largely uninteresting form of work). In its most incisive forms media archaeology sets out, using tactics including the counterfactual, precisely to question those *relations between* the past and present that have produced, as the present, and as the present imaginary, a set of naturalized objects and naturalized histories, already lined up and ranked *as* the 'mainstream'.

Jussi Parikka has taken up this issue, arguing that in his writing apparently unlikely, or trivial, or arbitrary object choices are valued because they can confound established *categories* through which standard histories work. Thinking of insects as media, for instance, he intends to confuse divisions between (what tend to be thought of as) media and other systems; precisely those divisions which set the boundaries of various forms of intelligibility – conditioning legibility in public discourse, or policing disciplinarity and knowledge claims between defined research fields, for instance. It can be argued that one way in which media archaeology *does* engage with 'the mainstream' is by challenging its border and limits, its banks and its dams.

Ernst's rationale for paying attention to technical media represents a different challenge to media archaeology's object choices. The burden of his position is that technical media constitute the mainstream even if this is not widely recognized (yet), nor its implications traced out. His objects of study are those determining the new overarching conditions of intelligibility. If computational technologies beat out our time, condition utterance, or discourse, and remake writing, then it is crucial that we study them, no matter how obscure they may appear. What is closely argued in Ernst is arguably present as an ambient sensibility informing digital media archaeology in general; it underpins claims made for the significance of the technical over the representational, a prioritizing that also determines how mediation is understood and what its stakes are. This is the case 'now' if not before, since 'now' we live in times of pervasive computing, in the aftermath of a shift in episteme ushered in *by* the rise of technical media. You don't have to go all the way with Ernst to recognize the value of the perspective his work opens up.

Media archaeology versus cultural studies?

Media archaeology has often been hostile to cultural studies. The latter is framed as preoccupied with representation, obsessing over the abstract operations of ideology, and focusing on the technological imaginary, rather than attending to the thing itself. The charge is of tilting at ghosts rather than tangling with the real, of failing to understand the impact, and importance, and/or the informing *force* of the material operations of technical media. There are commonalities between media archaeology and other engagements with revived forms of (new) materialism. The computational turn (see Wing, 2006, Lanier, 2010, Berry, 2011), traditional digital humanities, software, code, or protocol studies, objected-orientated philosophy (OOO), represent very different critical/post-critical orientations – but all resonate sympathetically with the media-archaeological suspicion of the utility of cultural studies approaches to the study of media (Bassett, 2020).

Against this I make a case for (a form of) media archaeology *as* (a form of) cultural studies, seeing this as both nascent in various ways and as a desirable evolution. The argument to be made is that cultural studies approaches can contribute to developing forms of more critical media archaeology, and media archaeology can contribute to re-materialization of cultural studies – and cultural histories. Bringing the two together produces new approaches; not least, it enables new forms of digital humanities to be developed. To stress, this is not a matter of joining what was before entirely divided; it is to acknowledge that media archaeology, like digital humanities, *is* a cultural study – or it is nothing.

Some unpacking is needed. This begins with an acknowledgement that in some of its iterations cultural studies *has* neglected the informing force of material in favour of an exclusive attention to representation, which latter is understood as powerfully performative, constructing that which it names in accordance with ideologically determined presuppositions and alignments. The result is a failure to attend to the operations of technical media and the forms of ordering that arise as a consequence of their technical affordances. If some forms of media archaeology strip out the imaginary as a necessary component of the technical, some cultural studies strip out the operations of – the matter of – the medium, producing a woefully flattened understanding of what is under investigation and of how it is powerful. Crude representationalism can never grapple with new forms of data-driven visuality that organize platform subjectivities, for instance, nor can it grapple with the space-time compressions/distensions characteristic of digital operations. However, currents within cultural studies have responded to the clash of materialisms, asking how historical materialism (and other modes of critical analysis) and the various forms of attention to 'the material' that new materialism wishes to prioritize can be brought into new relation. By the late 20th century, the *New Cultural Studies* reader (Birchall and Hall, 2006) had already included vociferous calls for new kinds of post-humanities – explicitly declaring the need to break with earlier traditions, or to remake them in relation to rising forms.

But cultural studies, certainly in its Birmingham or UK versions, never confined itself to representation, in any case. In many of its most influential forms it has been explicitly informed by a desire to engage with material culture. It is salutary to remember that Stuart Hall, thought of as a co-founder of the field, explored the racist discourses of policing in the UK through an exploration of 'mugging' which never lost sight of the material violence acted out on discriminated-against bodies (see Hall et al., 2013). Even Hall's work on encoding/decoding, widely viewed as the epitome of a particular kind of media and cultural studies, which avowedly *privileged* the semiotic moment in circuits of culture, did not set aside material processes and technologies involved in production, nor those involved in reception (Hall, 1992).

There are also connections to be made with a mode of anthropologically orientated media studies undertaken in the 1980s and 1990s, which evidences further an extant tradition of engagement between medium theoretic and more discourse-orientated research – and a prioritization of non-obvious research objects. Scholars of media and everyday life (Silverstone, 1994, Highmore, 2001) influenced by Hall found resources in the medium thinking of Marshall McLuhan and in Joshua Meyrowitz's (1986) exploration of technical time and space, as well as in the historically materialist work of Raymond Williams (2003). Of note is the degree to which the 20th-century French tradition of the study of everyday life informed this work – particularly that of Silverstone (1994) and Morley (2007). Following this trail is relevant because the French focus on the objects, textures, materials, and orderings of everyday life led to a re-evaluation of the significant. Perec's forensic investigation of the infra-ordinary, often deploying data-driven or automatic techniques to pay attention to apparently insignificant object (pens, desks, spoons), produces a meditation on everyday life and its objects that sees them as both what matters least and what matters most. It can be understood as simultaneously a kind of media archaeology and cultural studies (Perec, 2009, Bassett, 2017).

Perec's work can seem fey. Perhaps it appears consciously vogueish – to return to the critique of media archaeology invoked earlier. But a harder look at his writing on the everyday and its relentless rhythms and repetitions reveals an account that is always aware of matters of resistance, and also of systematic domination. At this point we can link it forwards and see it as part of that tradition of cultural studies which has never forsaken the material, but which has *also* always refused to divide the material from the symbolic, which has always been aware both of power and of its multiple operations; cultural Marxism, after all a reasonably standard description of Hall and the Birmingham School of Cultural Studies, was never purely a matter of the study of representations and the ideological positionings they revealed circulating in discourse.

Returning to terms Foucault used (and Wolfgang Ernst adapted), we might say that if cultural studies refuses the substitution of the document with the machine (Ernst's move), since it continues to maintain a sense of the importance of the symbolic – in the mediating operations of the computational, as well as elsewhere – it also refuses to stay with discourse. Rather, it enables a theoretical space within which *more* than a reversion to the document, as a response to the lacunae produced by media archaeology's focus on the purely technical as that which is significant about pervasive mediation, can be contemplated. To invoke an example here, in a later chapter I argue that the fact that ELIZA, a software program, could tell stories and had a cover story as a therapist was of significance. The argument is that these aspects of what ELIZA did were as materially informing in the bot's reception as the script and algorithm that formally constituted the program. Contra Ernst, machines may tell narrative histories. More, they may become operational in machinic ways. Going back to Foucault, and to the much-disavowed structuralism that, despite Foucault's own protestations, haunts the *Archaeology*, we might note that narrative itself has for a century or more been subject to attempts to understand it computationally (Bassett, 2007). ELIZA has also been explored through feminist theory (e.g. recently by Sarah Dillon, 2020) and another key locus here is feminism, particularly feminist techno-science, whose influence on cultural studies has been large and which has long understood and argued for a blurring of precisely these boundaries (e.g. Bassett, O'Riordan and Kember, 2020). Neither document nor technical media, then. Through cultural studies the Foucauldian document has been re-materialized differently – and in ways that *expand* rather than simply *switch over* what is taken as the proper object of study. The result is the possibility of a socio-technical media study containing multiple materials, multiple temporalities, multiple actors. This also defines an expanded – and desirable – form of digital humanities.

Forgetfulness and anti-computing

The basic position outlined above can now be further developed and brought closer to the proximate concerns of anti-computing through a consideration of forgetfulness – which latter can be understood as a feature of technical media, as a logic strongly operating within contemporary forms of computational capitalism, and as entailed in computational imaginaries. Forgetfulness is one key to understanding the operational version of the presentism identified as an extant feature of anti-computing. Accepting that divisions between media archaeology and cultural studies are not absolute, that parallels and shared orientations to be found, that the grounds each claim are already partly occupied by the other, then potentially they may be put to work to grapple further with what has been defined as a key characteristic of anti-computing, its persisting disappearance and reappearance across time, which is to say how it is forgotten and remembered. This dynamic challenges attempts to account for anti-computing through forms of history relying entirely on disjuncture, as well as those that understand significance and order strictly in terms of continuity and linearity. The issue might be how to bring these accounts into relation, and how that might be productive. Approaching this, I intend to work through forgetfulness from both sides.

First, then, technological forgetfulness and machine ontology. Contemporary emphases on technical media and the organization of temporal experience resonate with earlier work exploring cinema. The trajectory identified (roughly from Kittler to Ernst, from film to the computational) links the mechanization of the persistence of vision with the rise of computational operations and mediation, and sees both as the conditions enabling and organizing other social and cultural forms (which then become inadequate as ways of understanding the world, since they cannot capture the reality of things, including such media things – this is more or less Ernst's argument with narrative history). Kittler's discussion of cinema explores how technology automates the 'persistence' of vision, or the burning in of an image so that it persists as an after-effect, a phenomenon of human optics (Kittler 1999, Bassett, 2015). What is received is discontinuous but so organized that we may accept its results as continuous – to echo Foucault on the Panopticon, and by doing so point to how these arguments relate also to matters of power. In the case of the computational, persistence is not the issue, but a different form of discontinuous continuity *does* pertain: that which is produced by reprogrammability. The universal machine, based on reprogrammability and simulation, has to forget in order to operate the next program (and storage is not the same as working memory). Even neural network-based AIs find memory difficult.[6] Computational operations are discrete, even if, as Beatrice Fazi, suggests, there is a case for arguing that they are not entirely determinate (Fazi, 2018).

Instantiated computational technology expands this affordance to require forms of forgetfulness in its users, in particular, inviting, even demanding, forgetfulness (absent-mindedness) about the technical media system itself and about its mediating operations. This is the promise of direct connection, of object manipulation, of 'writing' on a screen without having to think about writing as coded. Here, then, the injunction to forget has taken a new form. It has it has passed *from* the necessary forgetfulness of the cut that joins, that also joins human vision to cinematic technique, to produce the image, which was essential to cinema. It has passed *to* an interface that conceals, and suggests forgetting the concealment not of a screen apparatus but, rather, of that which is not screened but is nonetheless computed and therefore mediated in new ways (not only as vision, but in code). A measure of what is deemed successful in interface design has, after all, long been that we (feel we) reach through, touch directly, *internalize* the interface (suddenly cinema and computing move closer together). An examination of the technical affordances of this form of technical media can suggest all this.

What is still missing and needs to be included is the degree to which forgetfulness is 'built in' as a standard industrial and marketing strategy. And this takes us towards matters of political economy. Consider, then, that the injunction to forget is intrinsic to what might be termed upgrade culture in general, the logics of which are exemplified in the upgrade itself. Upgrades, that is, are designed to ensure that sufficient continuity is quietly provided – whilst novelty is loudly delivered. An example at one scale is the tablet, emerging as an elision between the computer and the smartphone, but proclaimed as a new category; the versioning of software providing for high levels of skilled use by retaining key organizational features whilst promising radical advances constitutes another. The continuous renewal of the skeuomorphic features of interfaces and devices, a constantly upgrading invocation of what was there, in the last version, or the last but one, is characteristic of this formation in general; contrary to claims that the imitative interface is over (Worstall, 2019), imitation is a generalized principle. The chameleon capacities of the computer, and of code, its capacity to imitate, and the tendency to retain but push into the background what has become stable means computing developments can take on the feel of the era when they emerge, silently and very fast.[7] The result might be an amnesiac condition, partial but real, partly consented to, that could be said to extend from (what is given as the affordances of) technologies and platforms, through use, to users themselves. Contemporary generations of 'digital natives', supposedly the inheritors of an extreme form of what Jaron Lanier (2010) terms cultural neoteny, become relevant here. For these perpetually refreshed, and therefore perpetually infantile, groups of users, so we are often told, the world before the technological now, their digital, is literally 'unthinkable'. Moreover, their version is what counts. They are the mainstream, and the backwaters of the past, with all its old lags, human and technical, do not really count in the 'here and now' at all.

Finally, forgetfulness, explored through a consideration of the ontological qualities of the computational, of the contradictory demands for compatibility (familiarity) of the novelty which the market generates (and satisfies to some temporary extent), and of the kind of forgetfulness this encourages, may be explored at a larger scale still; that is, as organizing elements in the techno-social relations of computational capitalism. At this scale these relations entail another (related) kind of forgetfulness.

Compulsory technology?

Ellen Meiksins Wood has argued that an older market system became capitalist when it became compulsory (Wood, 2012: 40). Computing is now a core part of that (market) system's logics and underpinning operations, and, following Wood's line of thought, it can be argued that a variant on this – informational or computational capitalism – emerges as computation becomes necessary, or itself compulsory, as the figure and material form of that system, part of its emergence as a truly global system, organizing (the uneven terms of) relations for all within its spaces of flows and its backwaters (Castells, 1996).

We live within this formation. The global reach of satellites and the intimate touch of population databases, and the rise of biotech, provide proofs of that if they are needed. What does this mean? Does it mean that the *only* form of computing we have now is what we are 'given' by the market, and that the way of life (and forms of culture this enables/entails), the compulsory way of life, is the only form of life available? Is choosing between pre-set options the only choice that may be made? Matt Fuller has convincingly argued that the software forms which the market makes are not the only forms that may be made, and this pertains at other scales. I will return to this. For now, it is clear that what *is* implied is that computing arrives materially and symbolically in chains, that it *tends* to take forms or develop along trajectories that appear to be 'inevitable', or ontologically directed. It is easy, then, to *forget* or overlook the degree to which they are formed contingently in relation to markets or to the social structures that are bound into them. The forms of the computational are thus so mainstreamed that they appear 'natural', perhaps.[8] The computer industry itself plays a part in this. It forecloses, by way of the visions it offers, by way of the technologies it does not support, through the closing down of open systems, the systematic acquisition of the innovative by the giants, on that vision or sense of how computing could reach beyond the market and could take different forms entirely. It would rather we choose between those (highly constrained) 'choices' *within* the grounds it prescribes, and these are the grounds of compulsory capitalism.

This strongly articulated invitation to forgetfulness, discerned as part of the ontology of the computational object, identified as a material possibility exploited and operationalized by the computer industry in its designs, so that it becomes embedded, reinforced through the production of a technological imaginary that fetishizes the new but – in the manner of the fetish – also withholds it, can thus also be connected to dominant techno-cultural temporal logics with their intense focus on the present, and their claims, urgent environmental issues about the future notwithstanding, for the obliterating significance of the now. And we should talk of the *space* of the now as significant, since that prioritization of space over time, which Jameson (1991) understands as operationally dominant in late capitalism, is very clear here. Computational forgetfulness and present-obsessed capitalism are co-dependent in the contemporary formation, *designating* what constitutes its mainstream and informing its relation to the past and the future.

Anti-computing, as a found phenomenon, developed as a critical intervention and constituting a methodology, both confirms this state of affairs and disrupts it. It confirms it by its tendency to fade, to appear novel, to fall out of the present. It disrupts it firstly in that it points to the possibility of other possibilities by refusing or dissenting from that which is offered (and in that restricted sense it works by way of negation). Secondly, it is significant in that it arises at all, since in doing so it questions the authority that compels us forwards; *compulsion* falters. Anti-computational formations produce questions about what 'we' are supposed to want, or how 'we' are 'supposed' to think about computers, how 'we' are 'supposed' to forget them, on the one hand, and move on with them in the smoothly reassigned grounds of the permanent present of the markets they support and help to create, on the other.

A third way in which anti-computing is disruptive is that it arises *again*. Alongside forgetfulness, or the tendency of anti-computing formations to subside, to fail to have continuous purchase, to operate horizontally as it were, is that other (related) characteristic already identified – recurrence or recrudescence. When anti-computing arises or gains some purchase its forms are distinctive in that they relate to arising events or technological developments, but they are often also very familiar. The same questions have been raised, the same objections voiced, the same unease generated, around certain elements of computational culture, over and over again. Further, if they are often raised each time as if new and/or in less than perfect knowledge of the last time around, which is say that they are raised *forgetfully*, they also, having risen, reconnect and refind earlier forms of contestation. Anti-computing has a submarine legitimacy to draw upon.

Anti-computing is a recurring position, and the moments I explore, which exhibit complex temporalities within themselves and which are assembled as matter-centric collections, heterogeneous archives rather than documents, are fragments of a much larger-scale loose formation, operating across time and developing, scaling, albeit not in linear ways. The erratic persistence of many forms of anti-computing matters. To explore the form this takes demands a brief return to further explore the cyclicality of the reviving salience of anti-computational sentiment. This perhaps suggests a diversion, but the patterned recurrence characteristic of anti-computing formations suggests how anti-computing operates as the illegitimate other of compulsory computing and can also be used to clarify a distinction between compulsion (or what is compulsory) and (technological or any other) determination.

Valerie Traub's account of the recurring salience of bodily tropes over time can be reinvoked here. Salience is the term Traub uses to understand the recurrence of forms of bodily intelligibility across history, and thereby to account for a form of historical patterning. Traub is concerned to understand forms 'whose meanings recur intermittently and with a difference across time' (Traub, 2007: 126). She wants to understand *how* 'certain perennial logics and definitions remain useful, across time' and how these moments emerge 'at certain moments, silently disappearing from view … re-emerging as particularly relevant (or explosively volatile)' (Traub, 2007: 126). Traub is not talking about universal types or trans-historical categories that comprise or subsume historical variation, nor about basic concepts. Her focus is on something more contingent, and more likely to operate at multiple levels. She claims that the logics and definitions of the particular material forms bodies take 'tend to reappear in a different guise under changed social conditions', and notes that the discourses with which they are articulated shift and mutate as well (Traub, 2007: 128). In other words, reappearance is never total, and each emergence has singular, as well as common, qualities. This is not only a matter of meta concepts returning, but of a more heterogeneous persistence, what I might term a reactivation. What Traub wants to document (or takes as her 'document', relating this back to Foucault and 'doing history') is the technology of the flesh across long time spans. Hers looks like an account of the discourses that conform bodies, and so might be thought to be informed by the immaterial, to be 'only' a matter of representations and recurring discourses (in this sense it might seem to confirm the media-archaeological objection, and the long-standing objections to Butler's work on materialization, to which Barad responded with intra-activity theory (Butler, 1993, Barad, 2003). This reading of Traub forgets, though, that in her account it is through and in real bodies, *as real bodies*, that these returns are made – that conformation is productive and operates to shape matter. That is, it is bodies that operationalize and remake revenant discourses; bodies that are the grounds of this discourse, a key part of its fleshing out. To stress this obvious but often overlooked feminist insight – that bodies matter, which is what Butler also taught us – is also to point to an important way in which discourse and its materials are not to be entirely divided – and this may also be applied to the computational. As for bodies, intermixing the symbolic and discursive with the flesh, so for machines. The material is not all on one side and all the discourse and representation on the other. This is new materialism's fetish, but not one we need partake of.

Traub's argument constitutes an accounting with relativism and new history, an attempt to reintegrate a sense of continuity whilst avoiding a simple return to grand narratives. Drawing on archaeological approaches, she is interested in the limits of oscillation (deploying Foucault's term), pushing back against the tendency to develop forms of history that are entirely disjunctive; a tendency that has been identified earlier as a characteristic of some forms of media archaeology. Her avowed intention is to route around the binary set up by continuist and disjunctive theories of history (specifically, the history of sexuality, but the approach can be put to work here). This places Traub at some distance from the feminist 'anti-history' of Carla Freccero et al., which refuses all linear histories in favour of a kind of encounter (Freccero, 2006).

I go with that distancing, since what I want to generate is emphatically not (*pace* Freccero) an 'anti-history' of anti-computing. Nor, as is already evident, in so far as media archaeology refuses entirely a certain kind of narrative history, is what I want to do entirely media archaeological. The point is to let these moments of anti-computing come into relation with each other, to recognize their complex temporality, and also to think them through in relation to larger social fabrics and dynamics. Traub, though, has some sympathies with the 'encounter', even whilst she also breaks with it, and I too find this useful. She suggests a history 'motivated, both in form and content, by the question: how might we stage a dialogue between one … past and another?' (Traub, 2007: 137–138). This dialogue is not entirely 'invented' or 'staged' by the historian; the 'pasts' that are the subjects of this dialogue are *already* trespassing on each other's space, are already in dialogue. This is not a matter of a pristine emergence. Traub's thinking is informed at least as much by Benjamin's sense of non-linear time as Foucault's sense of history (Benjamin, 2006, Eagleton, 2015), and in this way also points us back to the influence of Benjamin on various forms of media archaeology itself. Her sense of the return of various representational features (Traub, 2007: 128), of bodies and bonds, is complex. She does not simply seek to bring the old back into view – this would bring back only the dead – nor to deal only with the found return of old representations, particularly if these are viewed as discrete from their ongoing and contingent incorporation (the parallels between bodily incorporation and machinic instantiation are clear here). Her concern is with what is brought back into active discourse and bodily operation, that is, *what is re-incorporated or re-materialized.* At issue is what becomes affective and active, and once again explosive; what may therefore also transform. Traub wants to ask what history does with its bodies. I find this interesting and helpful because I want to ask what it does with its machines. The continuous operations of markets both fragment and defragment technological histories by relentlessly bringing technology stories into the present (compressing them); only what is new and on the shelves can be supported (literally and metaphorically) by the continuous orderings and reorderings of neoliberalism, which include the reorderings of history. If this leaves the history of computational dissent in ruins, the point is not only to refind the ruins. It is to rethink how these ruins relate to other moments and to greater wholes. And it is also to see how shards of the past trespass on the present and might also in this way project into the future – perhaps as moments when compulsion is refused.

Coda

In these two opening chapters anti-computing has been developed and is used as an organizing concept. It takes the place of, for instance, a local focus, or a specialist history, or work around a single technology, moment, or subject. It tells a series of related stories and in doing so produces an overlapping ordering of things. It is the response I offer to Edwards' call to do computer history despite the difficulty of doing it. But anti-computing is, as well as an organizing perspective, something discerned out there in the world. It already *has* a certain coherence – the same forms and dynamics recur – and do so prior to the taxonomic operations of anti-computing as an identificatory methodology that I undertake so as to give them some further coherence, or to amplify their salience. The following chapters explore some unlikely, forgotten, revenant, familiar *and* strange features and moments of anti-computing.

Notes

1. *Datamation* is widely claimed as the first computer magazine.
2. See e.g. the Covert Criminal Intelligence Sources (Criminal Conduct) Act (2021).
3. Public and expert debates around image manipulation in the interests of truth and aesthetics continue in the 2020s, of course. See e.g. https://fstoppers.com/originals/thats-photoshopped-yeah-so-does-mean-all-our-photos-are-fake-451724 for a photographer's view.
4. See Ithiel de Sola Pool's *Technologies of Freedom* (1983) for an interesting example.
5. Chapter 4 of this book is in oblique dialogue with Huhtamo's serialization of interactivity, undertaken in relation to automation. I return to some of the same grounds but establish vectors that bring cybernation debates back into play in relation to labour-precarity issues rather than following his trail, which links cybernation to interactivity as a medium form (Huhtamo, 1999, see also Bassett and Roberts, 2020).
6. Their limited capacity to 'remember' is, however, something that differentiates them from other more forms of machine learning – for instance, Markov chains.
7. Catherynne M. Valente, *Silently and Very Fast* (2011).
8. See Graeber (2015).

Bibliography

Anthony, Scott. 2012. 'What is Media Archaeology?', *Reviews in History*, 1343, www.history.ac.uk/reviews/review/1343.

Badminton, Neil. 2006. 'Cultural Studies and the Posthumanities', in Gary Hall and Clare Birchall (eds), *New Cultural Studies: Adventures in Theory*. Edinburgh: Edinburgh University Press.

Barad, Karen. 2003. 'Posthumanist Performativity: Toward an Understanding of How Matter Comes to Matter', *Signs: Journal of Women in Culture and Society*, 28(3): 801–831.

Bassett, Caroline. 2007. *The Arc and the Machine: Narrative and New Media*. Manchester: Manchester University Press.

Bassett, Caroline. 2011. 'Twittering Machines', *Differences*, special issue 'A Sense of Sound', 22(2): 276–300.

Bassett, Caroline. 2015a. 'Not Now? Feminism, Technology, Post-digital', in Berry and Dieter (eds), *Postdigital Aesthetics: Art, Computation and Design*. London: Macmillan: 152–161.

Bassett, Caroline. 2015b. 'After Images of Cinema: Kittler and the Mobile Screen', in Stephen Sale and Laura Salisbury (eds) *Kittler Now: Current Perspectives in Kittler Studies*. London: Polity, 192–209.

Bassett, Caroline. 2017. 'What Perec Was Looking for: Notes on Automation, the Everyday, and Ethical Writing', in Rowan Wilken and Justin Clemens (eds) *The Afterlives of George Perec*. Edinburgh: Edinburgh University Press: 120–135.

Bassett, Caroline. 2020. 'Digital Materialism: Six Handed, Machine Assisted, Furious', in Peter Osborne (ed.) *Thinking Art: Materialism, Labours, Forms*.

Bassett, Caroline and Ben Roberts. 2020. 'Automation Now and Then: Automation Fevers, Anxieties and Utopias', *New Formations*, 98: 9–28.

Bassett, Caroline, Sarah Kember and Kate O'Riordan. 2020. *Furious: Technological Feminism and Digital Futures*. London: Pluto.

Beniger, James. 1989. *The Control Revolution, Technological and Economic Origins of the Information Society*. Cambridge, MA: Harvard University Press.

Benjamin, Walter. 2006. 'On the Concept of History', in *Selected Writings*, 4: *1938–1940*. London: Harvard University Press.

Berkeley, Edmund C. 1961. *Giant Brains: Or Machines that Think*. New York: Science Editions.

Berry, David M. 2011. 'The Computational Turn: Thinking about the Digital Humanities', *Culture Machine*, 12: 1–22.

Black, Edwin. 2011. *IBM and the Holocaust: The Strategic Alliance Between Nazi Germany and America's Most Powerful Corporation*. New York: Dialog Press.

Braybrooke, Kat and Tim Jordan. 2017. 'Genealogy, Culture and Technomyth: Decolonizing Western Information Technologies, from Open Source to the Maker Movement', *Digital Culture & Society* 3(1): 25–46.

Butler, Judith. 1993. *Bodies that Matter: On the Discursive Limits of 'Sex'*. London: Routledge.

Castells, Manuel, 1996. *The Information Age: Economy, Society and Culture, I The Rise of the Network Society*. Oxford: Blackwell.

Ceruzzi, Paul. 1999. 'Inventing Personal Computing', in Donald MacKenzie and Judy Wajcman (eds), *The Social Shaping of Technology*. Buckingham: Open University, 64–86.
Chun, Wendy Hui Kyong. 2008. 'Sourcery or Code as Fetish', *Configurations*, 16: 299–324.
Chun, Wendy Hui Kyong. 2008a. 'The Enduring Ephemeral, or the Future Is a Memory', *Critical Inquiry* 35: 148–171.
Chun, Wendy Hui Kyong. 2008b. 'Programmability', in Matt Fuller (ed.) *Software Studies: A Lexicon*. London: MIT Press, 224–228.
Couldry, Nick. 2014. 'Inaugural: A Necessary Disenchantment: Myth, Agency and Injustice in a Digital World', *Sociological Review*, 64(4): 880–897.
Cox, Kevin R. 2001. 'Reviewed Work: The Origin of Capitalism by Ellen Meiksins Wood', *Annals of the Association of American Geographers*, 91(1): 220–222.
Cramer, Florian. 2014. 'What is "Post-digital"?' *APRJA Post-Digital Research*, 3(1).
De Sola Pool, Ithiel. 1983. *Technologies of Freedom*. London: Harvard University Press.
Dillon, Sarah. 2020. 'The Eliza Effect and Its Dangers: From Demystification to Gender Critique', *Journal for Cultural Research*, 24(1): 1–15.
Eagleton, Terry. 2015. *Hope without Optimism*. New Haven: Yale University Press.
Edwards, N. Paul. 2001. 'Think Piece – Making History: New Directions in Computer Historiography', *IEEE Annals of the History of Computing*, 23(1): 88.
Elsaesser, Thomas. 2004. 'The New Film History as Media Archaeology', *Cinémas*, 14(2–3): 75–117.
Elsaesser, Thomas. 2016. 'Media Archaeology as Symptom', *New Review of Film and Television Studies*, 14(2): 181–215.
Ernst, Wolfgang. 2000. *M.edium F.oucault. Weimarer Vorlesungen über Archive, Archäologie, Monumente und Medien*. Weimar: VDG.
Ernst, Wolfgang. 2013. *Digital Memory and the Archive*. London: Minnesota University Press.
Fazi, M. Beatrice. 2018. *Contingent computation: Abstraction, Experience, and Indeterminacy in Computational Aesthetics*. London: Rowman & Littlefield International.
Foucault, Michel, 2001. *The Order of Things: Archaeology of the Human Sciences*. London: Routledge.
Foucault, Michel. 1972. *The Archaeology of Knowledge*. London: Routledge.
Foucault, Michel. 1980. *Power/Knowledge*. London: Harvester Press.
Foucault, Michel. 1997. 'What is Critique?', in Sylvère Lotringer and Lysa Hochroth (eds) *The Politics of Truth*. New York: Semiotext(e.).
Freccero, Carla. 2006. 'Queer/Early/Modern', *Series Q*. Durham, NC: Duke University Press.
Fuller, Matthew. 2008. *Software Studies: A Lexicon*. London: MIT Press.

Gerovitch, Slava. 2001. '"Mathematical Machines" of the Cold War: Soviet Computing, American Cybernetics and Ideological Disputes in the Early 1950s' *Social Studies of Science*, special issue 'Science in the Cold War', (31)2,: 253–287.
Gitelman, Lisa. 2006. *Always already New: Media, History, and the Data of Culture*. London: MIT Press.
Graeber, David. 2015. *On Technology, Stupidity and the Secret Joys of Bureaucracy*. London: Melville House.
Haddon, Leslie. 1988. 'The Home Computer: The Making of a Consumer Electronic', *Science as Culture*, 2: 7–51.
Hall, Gary and Clare Birchall (eds). 2006. *New Cultural Studies: Adventures in Theory*. Edinburgh: Edinburgh University Press.
Hall, Stuart. 1992. 'Encoding/Decoding', in *Culture, Media and Language*. London: Routledge, 128–138.
Hall, Stuart, Chas Critcher, Tony Jefferson, John Clarke and Brian Roberts. 2013. *Policing the Crisis: Mugging, the State and Law and Order (2nd edn)*. Basingstoke: Palgrave Macmillan.
Hayles, N. Katherine. 2008. *Electronic Literature: New Horizons for the Literary*. Notre Dame: University of Notre Dame Press.
Helmore, Edward. 2000. 'Be Very Afraid of the Future', *The Observer*, 19 March, www.theguardian.com/theobserver/2000/mar/19/focus.news1.
Hendy, David. 2000. 'A Political Economy of Radio in the Digital Age', *Journal of Radio Studies*, 7(1): 213–234.
Hertz, Garnet. 2010. 'Archaeologies of Media Art, Jussi Parikka in conversation with Garnet Hertz', *CTheory* (Spring), www.ctheory.net/articles.aspx?id=631.
Highmore, Ben. 2001. *The Everyday Life Reader*. London: Routledge.
Huhtamo, Erkki. 1999. 'From Cybernation to Interaction: A Contribution to an Archaeology of Interactivity', in Peter Lunenfeld (ed.) *The Digital Dialectic. New Essays on New Media*. Cambridge, MA: MIT Press, 96–110.
Jameson, Frederic. 1991. *Postmodernism, or, The Cultural Logic of Late Capitalism*. Durham, NC: Duke University Press.
Jameson, Frederic. 2015. 'The Aesthetics of Singularity', *New Left Review*, 92: 101–132.
Kittler, Friedrich. 1997. *Literature, Media, Information Systems*. London: Routledge.
Kittler, Friedrich. 1999. *Gramophone, Film, Typewriter*. Stanford, CA: Stanford University Press,
Kittler, Friedrich. 2008. 'Code (or, How You Can Write Something Differently)', in Matthew Fuller (ed.) *Software Studies: A Lexicon*. London: MIT Press, 40–48.
Lacey, Kate. 2013. *Listening Publics*. Cambridge: Polity.
Lanier, Jaron. 2010. *You are Not a Gadget*. New York: Knopf.
Lanier, Jaron. 2012. 'One Half of a Manifesto', *Wired*, 8(12), http://wired.com/archive/8.12/lanier_pr.html.

Levy, Steven. 1994. *Insanely Great*. London: Penguin.
Light, Ann. 2011. 'HCI as Heterodoxy: Technologies of Identity and the Queering of Interaction with Computers', *Interacting with Computers*, 23(5): 430–438.
Liu, Alan. 2004. *The Laws of Cool: Knowledge Work and the Culture of Information*. Chicago: University of Chicago Press.
Lovink, Geert. 2003. 'Interview with German media archeologist Wolfgang Ernst', *nettime*, 25 February, www.nettime.org/Lists-Archives/nettime-l-0302/msg00132.html (accessed September 2015).
McLuhan, Marshall. 2002. *Gutenberg Galaxy*. Toronto: Toronto University Press.
Meyrowitz, Joshua. 1986. *No Sense of Place*. Oxford: Oxford University Press.
Miéville, China. 2009. *The City and the City*. London: Pan Macmillan.
Morley, David. 2007. *Media, Modernity and Technology: The Geography of the New*. London: Routledge.
O'Riordan, Kate. 2018. *Unreal objects*. London: Pluto Press.
Parikka, Jussi. 2013. 'Archival Media Theory: An introduction to Wolfgang Ernst's Media Archaeology', in Wolfgang Ernst, *Digital Memory and the Archive*. London: University of Minnesota Press, 1–23.
Parikka, Jussi. 2010. *Insect Media: An Archaeology of Animals and Technology*. London: University of Minnesota Press.
Parikka, Jussi. 2012. *What is Media Archaeology?* London: Polity Press.
Perec, Georges. 2009. 'Notes on the Objects to be Found on My Desk', in *Thoughts of Sorts*. London: David R. Godine.
Roberts, Ben. 2020. *Critical Theory and Contemporary Technology*. Manchester: Manchester University Press.
Roberts, Ben and Mark Goodall (eds). 2019. *New Media Archaeologies*. Amsterdam: Amsterdam University Press.
Silverstone, Roger. 1994. *Television and Everyday Life*. London: Routledge.
Skågeby, Jörgen and Lina Rahm. 2018. 'What Is Feminist Media Archaeology?', *communication +1*, 7(1), https://scholarworks.umass.edu/cpo/vol7/iss1/7 (no pagination).
Solnit, Rebecca. 2007. *Storming the Gates of Paradise*. Cambridge: Cambridge University Press.
Steyerl, Hito. 2013. 'Too Much World: Is the Internet Dead?', *e-flux*, 49.
Stiegler, Bernard. 2013. *What Makes Life Worth Living: On Pharmacology*, Oxford: Polity.
Traub, Valerie. 2007. 'The Present Future of Lesbian Historiography', in George E. Haggerty and Molly McGarry (eds) *The Companion to Lesbian, Gay, Bisexual, Transgender, and Queer Studies*. Oxford: Blackwell, 124–145.
Turner, Fred. 2006. *From Counterculture to Cyberculture: Stuart Brand, the Whole Earth Network, and the Rise of Digital Utopianism*. Chicago: Chicago University Press.

Uprichard, Emma. 2013. 'Focus: Big Data, Little Questions?' in *Discover*, 1, http://discoversociety.org/wp-content/uploads/2013/10/DS_Big-Data.pdf.
Valente, Catherynne M. 2011. *Silently and Very Fast*. Washington, DC: WSFA Press.
Virlio, Paul. 1997. *Open Sky*. London: Verso.
Whitsitt, Samuel P. 2013. *Metonymy, Synecdoche, and the Disorders of Contiguity*. Padua: Libreria universitaria.it.
Williams, Alex and Nick Srnicek. 2013. '#ACCELERATE MANIFESTO for an Accelerationist Politics', *Critical Legal Thinking*, 14 May.
Williams, Raymond. 2003. *Television, Technology and Cultural Form*. London: Routledge.
Wing, Jeanette M. 2006. 'Computational Thinking', *Communications of the ACM*, 49(3): 33–35.
Wood, Ellen Meiksins. 1999. *The Origin of Capitalism*, New York: Monthly Review Press.
Wood, Ellen Meiksins. 2012. *The Ellen Meiksins Wood Reader*, ed. Larry Patriquin. London: Brill.
Worstall, Tim. 2019. 'Apple's iOS7, Well, It Was Time For Skeuomorphism To Die', *Forbes.com*, 19 September, www.forbes.com/sites/timworstall/2013/09/19/apples-ios7-well-it-was-time-for-skeuomorphism-to-die/#2ee8f58b455a (accessed February 2020).
Zielinski, Siegfried. 2006. *Deep Time of the Media, Toward an Archaeology of Hearing and Seeing by Technical Means*. London: MIT.
Zuckerberg, Mark. 2019. 'A Privacy-Focused Vision for Social Networking', 6 March, www.facebook.com/notes/mark-zuckerberg/a-privacy-focused-vision-for-social-networking/10156700570096634/ (accessed January 2020).

Chapter 3

A most political performance: treachery, the archive, and the database

This chapter undertakes a medium-theoretic analysis of the life of Harvey Matusow, Communist Party member, McCarthyite informer, and a man who famously recanted. Matusow described his early betrayals in terms of a need for recognition in a world of spectacle and in terms of automation – the camera and its distancing vision making the act of informing on his previous allies palatable. In later life he became a vocal opponent of computers and of the database society, founding an anti-computing league to fight against the tyranny of the automated sort and the automated cache. At one point he claimed as many league members as there were computers in England. Drawing on papers from the Matusow archive, this chapter tells an anti-computing story with medium transformation, mediatization, and the politics of identity at its heart. It plays into the present as an early iteration of database anxiety; and it haunts partly because it foreshadows the dangerous mixture of ignorance, incompetence, and authoritarian malice that characterized dealings around the Snowden events.[1]

* * *

> I don't like fascism, don't like bureaucracy, I don't like technology. (Harvey Matusow, interview, cited in Berenyi, 1971)

> A witness is a paradigm case of a medium: the means by which experience is supplied to others who lack the original. (Durham Peters, 2001: 709)

This is the outrageous story of Harvey Matusow, notorious and later notoriously repentant anti-communist, House Un-American

Activities Committee (HUAC) witness, campaigner for Senator McCarthy, contributor to counter-cultural magazines *Oz* and *International Times* (*IT*), and founder of an anti-computer campaign – the International Society for the Abolition of Data Processing Machines – which flourished in the late 1960s and early 1970s, largely in the UK, at one point claiming as many members in England as there were computers.[2]

Rolling Stone magazine described Matusow as 'Hustler Supreme', a man who operated under a series of names, boasting that he had 'led twelve lives' (*Rolling Stone*, 1972: 17 August). His early adulthood was in the Cold War climate of the US. His journey was from US Communist Party (CPUSA) member, to informer and professional blacklister, to anti-database campaigner, via a term in prison, a period in exile, a portfolio career of bewildering dimensions, and a string of personal reinventions; a life evidencing complex relationships between identity, publicity – and new forms of data sort and storage.

Matusow is remembered today in relation to the public anti-Communist hearings, but his life story is better understood in relation not only to spectacle but to the list, the archive, and the database. Operating as a professional anti-communist witness in the early 1950s Matusow destroyed the careers and lives of many former Communist Party friends and associates, and those of hundreds, if not thousands, of other people. He did so by dealing in information – generating and collecting it, witnessing to its veracity, and *claiming* veracity for it through the force of his invocation of it. Matusow gave up, and made up, names, contributing to blacklists, and invoking names in public, notably in the courtroom. Writing of the parallels between the trials and the McCarthy witch-hunts, the playwright Arthur Miller invokes the concept of 'spectral evidence', important in the events that inspired *The Crucible*. This required 'not the accused person, but [only] his familiar spirit' to be found committing a crime (Miller, 2000). Matusow's 'spectral' identifications' (Communist by implication, association, innuendo) assumed substantial, persistent, and damaging form – and had disastrously real effects.

Following a recantation, Matusow found a role as an ambiguous, and not necessarily welcomed, godfather figure in the UK counter-culture. He was involved in the London Film Co-Op, and became increasingly concerned with computers, then embedding in business and industry, powering government data banks, surfacing in various realms of everyday life, and clearly about to expand their reach further and faster, threatening, as Matusow saw it, 'to swallow us whole' (Matusow, 1968a).[3] In many ways an untaught medium theorist – and avowedly influenced by McLuhan – Matusow understood computers as media and information systems, machines for the storage and transmission of data, including personal data, which might eventually encode the entire world. And he hated them.

The result was the launch of the International Society for the Abolition of Data Processing Machines (ISADPM) – President, Harvey Matusow. The society was in many ways a media construct and little substantial activity took place on the ground, although much was promised. It was a minor, but for a time significant, part of a broader public debate in England around the emerging database society. It resonated with a cultural unease around computerization and its medium effects felt in the late 1960s, what might be termed a database anxiety. For many at that time computers seemed alien and the proposition that they might be 'far more deadly than the Beatles' yellow submarine', as one US Congressman put it,[4] not entirely absurd.

There is something bathetic about the later lives of Harvey Matusow. In the 1950s he was at the heart of events of global significance in 20th-century history, deeply entangled in the US domestic response to the Cold War. By the 1960s he was a minor figure in the UK counter-culture, chairman of a tiny organization in many ways as unreal as the front organizations he had operated for in the 1950s, and certainly less influential. Matusow's own unstable and unreliable personality encourages schismatic readings of his life. He was coy, often frankly revisionist, about his past (e.g. *Rolling Stone*, 1972). My proposition, however, is that the lives he led were not entirely disconnected, and that tracing out connections between them can be productive.

Undertaking this task, this chapter has two aims. The first is to explore questions concerning identity arising in relation to databases and their automation, where identity is understood as that which pertains to the self, and where the latter (computerized databases) are explored in terms of the processes they enable, including those that remake the ways in which the individual is known to, defined by, produced by, and controlled by various others – including the state. Matusow himself came to believe that database automation posed a threat to identity, dignity, and freedom. As he saw it fascism, bureaucracy, and (computer) technology were allied. His own experiences suggest why. This is not a psychological reading, however. Rather, Matusow's story is understood as symptomatic of broader cultural formations. The moment of the anti-computing league coincides with (contributes to) rising awareness of the social consequences of automation in the UK, and specifically to anxiety around the computerization of everyday life, and perhaps of 'life' itself.

The second aim emerges with the realization that this is a medium history. It is this both in terms of its content – Matusow, like McLuhan, was an early media operator – and because it challenges overly linear genealogies of media history that emphasize, to the exclusion of all else, a shift from representation (old technology) to information (new technology). Matusow's own life – let us take it as a medium and make its examination a method – points to more complex entanglements and trajectories. His blacklisting activities constitute a forcible reminder that the database flourished in other forms before the advent of its computerized form. New media(tion) is (also) old. This can be pushed further. The film theorist Thomas Elsaesser argued that the internalization of elements of 'cinematic perception' had put us 'in … the cinema' by the middle of the 20th century, but also insisted on cinema's entanglement within a broader audiovisual history including the phonograph and perhaps the Babbage engine (Elsaesser, 2004: 76). Elsaesser's material history provides an oblique commentary on user histories, and Matusow's story suggests how these could be extended. For Matusow, at least, questions of storage and access underpinning technical media were integral to the *cinematic* spectacle as he experienced it 'live', even if they were not (yet) internalized as more generally extant modes of cognition. More continuities emerge if this is viewed the other way around; the spectacularization of the political sphere entailed by McCarthyism, emphasized by the cinematic quality of the HUAC hearings, *remains* a key mode of politics in today's *computational* times. It is, albeit in altered form, a key feature of contemporary network media politics. Jodi Dean (2001), amongst others, has explored the queasy combination of abundant but terminally self-referring contributions that constitutes network media in communicative capitalism, an ecology that supports endlessly circulating forms of spectral performance, not least by ensuring their *performative* power.

The archive

Matusow travelled, but much concerning his life is collected in one place. The Matusow repository at the Special Collections unit at the University of Sussex includes papers and personal documents assembled by the man himself. The collection is split, roughly dividing the early and late lives. Each section consists of a series of document boxes, along with a small collection of books (e.g. last *Whole Earth Catalogue*, writings on communism in the US, the autobiographic *False Witness*) and other visual and audio-visual material, including *The Stringless Yoyo*, a self-made surrealist documentary made in 1961. Exploring the collection, I have, not without compunction, accepted Matusow as the archivist of his own life. To open the brown cardboard boxes containing materials he assembled is to feel his ghost at my shoulder; there are page numbers scrawled on the front of magazines (e.g. *New Scientist*),[5] envelopes with notations clarifying addresses, times, post-coding events, marginalia in the book collection. There is also what he left out.

This collection is – of course – a database of a kind, and the connections between fear of a database society powered by computers and controlling identity, a distrust of mainstream media arising out of an intimate knowledge of its operations, and the desire to send an account of oneself into the future by organizing the materials of your own memorialization are easy to see. Matusow, who knew lists could produce lies as truths and make those 'truths' operational, in the end bundled up the analogue materials of his own life story, including books, papers, stock photography, clippings, original documents, short-playing records – for posterity. Perhaps he did so in a bid to protect his name from the kinds of automated categorizations he campaigned against in the anti-computing campaigns and from the kinds of distortions and untruths (character assassination) he himself promulgated through his spectral identifications. It is thus ironic that parts of the Matusow collection are digitalized and in that process are de/recontextualized, in the name of access. I wonder how Matusow himself, who gave his story, and specifically those sections on prison and the McCarthy era to the 'Librarian and the Head of Department of American Civilization'[6] at Sussex and whose last will and testament document states that 'final authority' 'for any question which may arise from the use of these files … be the Librarian of the University and the head of Department of Department of American Civilization jointly',[7] would take this? Whilst he did wish that the material 'be available for research without restriction', digitalization involves the liquidation of the reassuring bulk of the cardboard on the original passport, the ink on the notebook, the fixer on the black-and-white 10 x 8 photographs, and the insertion of his history, now in coded form, into far vaster databases stored on the computer technology that, at one point at least, he disliked so much.

Looking at the detritus of a life,[8] I am enrolled as both voyeur and collaborator, engaging in the dissection *and* the reproduction of an endlessly fragile ego. Further, it appears that this is an ego that first consciously felt itself at risk not from information technology but in relation to the kind of valorization of spectacle that makes only one kind of life real or meaningful – that which is displayed in public and celebrated. The collection signals across the years an obsession with identity, the stable establishment of which appears at once to be desired and desperately to be avoided. A word that recurs endlessly – on greeting cards, on a pass signed by 'Joe McCarthy', in articles, plays, recantations, dubious legal documents, letters home, remission certificates, a series of passports, society photographs, hotel postcards, a 45rpm record – is 'Harvey'. Albert E. Kahn, the radical lawyer who dealt with Matusow's recantation, said '[h]e had apparently been unable to discard anything mentioning his name' (Kahn, 1955: 48). So perhaps Matusow's fear of computerization concerned not only what might be retained as 'fact' and regarded as such, but what might be lost. Pre-capitulating concern with the 'enduring ephemeral' (Chun, 2008) of digital memory forms, Matusow feared that automation would amplify what was already a faulty ingestion process; memory itself being lossy.

What can be learned from somebody who has revealed himself as a liar? In the preface to Matusow's autobiography, *False Witness*, which he helped to produce, Kahn asks '[h]ow can one be sure that … having lied so profusely … [Matusow] is now actually telling the truth?' (Matusow, 1955: 14). The latter addressed his 1950s readers directly: 'I know that many people will wonder how they can believe me now, when I have lied so often in the past; and I do not expect to be taken merely at my word. Readers will have to judge the truth for themselves' (Matusow, 1955: 17). The archive material in Special Collections could be viewed as another invitation to people to judge 'for themselves'.[9] Something raised by the Matusow case, after the 1950s, and the 1960s, and after the 1990s web with its endearingly hopeful sense of possible completeness, is the relation between the archive and the archived, and the distance between an archive as *constituted* and the new (increasingly computational or differently computational) forms it may take as it *persists*. Into this is inserted the question of the relationships and determinations arising across the archival process, particularly relations involving the depositor, and the archive's human and non-human curators. Matusow wanted the cardboard and celluloid archive he sent into posterity to speak his always contingent, always unreliable 'truth'. What follows opens the archive.

Identity

Matusow was born in the Bronx in New York on 3 October 1926. His early life was painfully unremarkable, as is made clear from the collected personal effects. Amongst them a Bar Mitzvah card, dated 1939, reads: 'Congratulations on achieving this first milestone to a healthy and happy manhood'.[10] A wartime ration book shows that by the age of sixteen Matusow (sex: 'M', weight: 160lbs, height: 5ft 8in, occupation: 'School'), had a number; 713043-FB. He kept his parents' ration books too, amongst the official documents he hoarded. He did badly at school, graduating on the basis of enlistment (Lichtman and Cohen, 2004). A yearbook for Taft Senior in 1945 lists him as in the armed forces.[11] He fought in the Second World War, spending two years (1944–46) with the US army in Western Europe, writing diligently, but with apparently very little to say, to his parents. He returned to New York and joined the Communist Party (CP) in 1947 and was initially an active member, including in various petitioning operations. In a filmed interview with Jean Luc Godard, Matusow said he had joined the CP to be 'somebody' once again, rather than being part of a crowd. He told Goddard that he was too impatient to work on writing, which might have given him a name. Instead, as he put it in *False Witness*, 'wanting identity', 'I chose the short cut, I joined the Communist Party' (Matusow, 1955: 21, 26).[12] Matusow's period of active membership of the CPUSA was short, although not untypically so, according to Ernst and Loth's (highly ideological) contemporary profile of the CP 'everyman' (Ernst and Loth, 1952).[13] He was for a time very active, but was expelled from the CP in 1951,[14] accused of being an 'enemy agent' (Matusow, 1955: 33, *Daily Worker*, 19 January 1951)· This accusation was justified, since by February 1950 he had contacted the Federal Bureau of Investigation (FBI) (Matusow, 1955: 27), to say he wanted to inform. Another milestone; not one that would lead to happiness.

Collecting

> Each of the 144 clicks of my shutter caught the face of a friend ... [This was] my first major report ... My future as an informer depended on the success of my picture taking. I overcame the hesitancy which I had in taking my first picture with my second picture ... the deed had been done ... although these were my friends ... it seemed almost impersonal to me. The camera was functioning as an informer, not I. (Matusow, 1955: 30, 31–32).

On May Day 1950 Matusow attended a New York parade as an FBI informer. There he 'took photos of my friends' (Matusow, 1955: 31). With this move he became a collector of names, a contributor to the linked and discrete databases gathered as part of the anti-communist operations of the US government that began from 1947, before the McCarthy era proper (Cameron, 1987), and ran on well into the 1960s. These operations included loyalty programmes, congressional hearings including the HUAC hearings, blacklisting – most notoriously of those in the entertainment industry but also of those in journalistic and university professions, trade unionists and industrial activists, and the use of various legal instruments including the Taft Hartley and the Smith Acts (Cameron, 1987).

Through his informing camera, which bought him more distance from his sense of personal betrayal with every shot, Matusow became an actor in a greater machine, a larger apparatus – one that often appeared to him to be as impersonal as the camera. Moving from photography 'for his friends' (who sometimes *asked* him to photograph them), to image capture as 'evidence' for the blacklisters, Matusow reworks photography from image to data.

Matusow's activities as a paid informer, an industrial spy, and a professional witness, have been considered by many historians working within the context of broader histories of aspects of McCarthyism. Lichtman and Cohen (2004), who used the Matusow Collection to produce an authoritative account of the 'deadly farce' that constituted Matusow's engagement with the American establishment during the McCarthy era map this scholarship, and I draw on them here along with the evidence in the collection. The essential facts are these. Matusow informed and testified for the FBI in HUAC sessions, gave evidence to Justice Department cases in various court and hearing sessions, and also directly campaigned for McCarthy's Senate re-election campaign on an anti-communist platform. He was active as a professional spy and engaged with private blacklisting and anti-communist publishing outfits, notably Counterattack (see later).

What began that May 1950 grew into a cluster of related activities that together constituted the working life of a professional anti-communist expert. Others were there before him; Matusow placed informers including Elizabeth Bentley, Matt Cvetic,[15] Louis Budenz, and Herbert Philbrick at the top rung of his new profession. His own 'career' began with a low-key first appearance in a closed session of HUAC, in which he was at pains to appear modest, not wanting 'to be seen as a glory seeker'. Kahn notes that, modest or not, his testimony was reported in sensational terms (Khan, 1955: 44). In 1952 Matusow gave open testimony to HUAC after being subpoenaed to do so. A letter to his parents documents his excitement about his more prominent role in the spectacle:

'Dear Folks'

well we had the first snow of the season ...

... don't say anything ******* both of you ... on 27th November I am going to start testifying at the Capital before the House Committee on UnAmerican Activities ... I have been officially called and am very happy about it ... DON'T SAY ANYTHING TO ANYBODY ... (2 November 1951).[16]

He was congratulated for this appearance in many quarters, including in *Counterattack* (Matusow, 1955: 43), which later made him an associate editor. By this time any diffidence about the scale of claims made, or the names given, had gone. Matusow later described himself as then a 'publicity addict', adding that 'to see the headlines scream and a few people cringe' was satisfying, and so was intervening in news agendas; being 'able to sell a story gave me a badge of importance' (Matusow, 1955: 65–67). The November 1951 letter concludes with this: 'When it happens you'll see me all over the NY newspapers, when this happens save them ... That's all for now ... Love Harvey.' The rewards of informing were material, providing earnings and entrée into particular kinds of society, but were also measured by Matusow in terms of fame: 'I was addicted to what printers' ink could do' (Matusow, 1955: 70). The archive itself is evidence of Matusow's fascination with the inky spectacle of himself, containing many assiduously collected clippings from this period.

Witnessing?

John Durham Peters defines witnessing as 'a paradigm case of a medium: the means by which experience is supplied to others who lack the original' (Durham Peters, 2001: 709). False witness is defined in the book of Exodus as raising a false report.[17] Given that Matusow very often lacked original information and that the names and accounts he gave were culled from other press reports, reliant on fabricated connections, or made up, he was from the start a false witness. But false or not, it was 'witnessing' that was key to his burgeoning activities. As he recognized, he did not become a witness *because* of his expertise in communism but, rather, used his 'reputation as a witness to establish myself *as* an "expert" on Communism' (Matusow, 1955: 82). With Bentley and the rest he became known as a former communist turned testifier, somebody uniquely qualified to speak about CP activities.

Matusow had a specialism. Youth was his 'gimmick' (Matusow, 1955: 68). He was introduced in the media as an 'authority on the

Communist conspiracy to penetrate the youth of America' (Kahn, 1955: 11).[18] His 'revelations' included the CP's use of intellectual and sexual lures as modes of indoctrination (Matusow, 1955: 77), its sinister rewriting of nursery rhymes, Red Scouts, and lurid reports on licentious goings on at a CP summer camp on Lenin's Rock. Also notable was Matusow's activity around New York schools; a Clark H. Getts presentation leaflet describes him as formerly a 'leader of the Kremlin's youth movement in this country', adding that 'amongst his sensational revelations in the press and before Congressional committees was the fact that there have been "more than one million card-carrying Communists in America, and more than 3500 young communists in the schools of New York alone"'.[19] He was also involved in a thwarted attempt to designate Antioch College a red base, and in actions over 'suspect' books held in State Department overseas libraries (see also Zinn, 1980).

In these and other activities testimony itself became evidence of expertise and expertise reinforced testimony. Matusow's time as an FBI informer was brief, since the FBI quickly understood that his claims were unreliable, but the Justice Department was happy to use him as a witness because of his HUAC reputation as an FBI informer. In this activity, which brought him into contact with Roy Cohn, assistant US attorney for the Rosenberg case, and still notorious, Matusow gave Grand Jury testimony in the Clinton Jencks case (Matusow, 1955: 190). Jencks, a union leader, was indicted and sent to prison for offences against the Taft Hartley Act. Matusow thereby arrived at the pinnacle/nadir of his professional anti-communist activity; the Jencks case was the linchpin in his later recantation. But there were many other activities before that recantation, since, in the public world of politics, there were many things that witnessing qualified him for. In 1952 Matusow was an aide in McCarthy's bid for re-election. He spoke on anti-communism in various states. His personal life was transformed. On the way up the ladder as a witness he married Arvilla Bentley, an heiress active in McCarthyite circles. As a moneyed celebrity he dined out, often in the very best places, on his fame and reputation, and on the misery of those he had incriminated. The menus and the photographs show up in the boxes.[20]

Lists and databases

Dear Folks, ... Please send me a copy of the "Worker" ... DON'T FORGET. It's important that I get it. Love Harvey. (Letter from Matusow, American Red Cross, 8 October 1951)[21]

The spectacle of the staged hearings is emblematic of the era in popular culture. But supporting the spectacle were the subterranean activities of collection and listing; the naming, capturing, storing, and use of names. Matusow was heavily engaged with this. Between hearings he briefly became 'official communist hunter for the State of Ohio' (Matusow, 1955: 55)[22] and an industrial spy. Back in New York, he associated with anti-red journals and with their blacklisting operations. One of these was Counterattack, an organization combining journalism (including the eponymous journal) with blacklisting – the latter often undertaken as free enterprise. Counterattack's owners were American Business Consultants (ABC), a firm offering 'security' information to employers, including General Electric and many large department stores in New York (Matusow, 1955: 109). Cold War historian Ellen Schrecker notes that whilst ABC were known for *Red Channels*, the entertainment blacklist, their activities were at least as expansive in other areas (Schrecker, 1998).[23] Counterattack was also involved in the murky and lucrative business of 'clearing' smeared names (Lichtman and Cohen, 2004, Matusow, 1955, Schrecker, 1998). Matusow sold subscriptions to the print publication (something he must have felt at home with, given his time in the CP), but this was only the visible tip of the larger operation. Counterattack *ran* on lists. Matusow describes 'thousands upon thousands of names ... cross-indexed with corresponding full documentation' (Matusow, 1955: 111). McCarthyism undeniably operated through mass media and spectacle, and Matusow himself described McCarthy as 'America's first electronic demagogue, riding the airwaves to preach his cause' (Matusow, 1969), but the listing operations underpinned the public show and extended far beyond it.

The dangerous connections between these categories are evidenced in Matusow's own activities, only initially to be understood as divided between collecting and testifying. Lists invoked in media accounts – accepted as really existing lists, independently of the specific entries that made them up – themselves became guarantors of proof when invoked in the sensationalized reports to which they lent some credence; this despite the fact that they were arbitrarily gathered, and often themselves relied on earlier press reports for their construction. Evidence of this lethal circularity can be found in the terrible and banal ways in which teachers were entrapped, and in the notorious claim made to a congressional hearing that there were 'well over 100 dues-paying Communists on the staff of the Sunday edition of the *New York Times*'. The total staff of the publication numbered far less than that, but nonetheless the claims were not entirely set aside (Kahn, 1987: 11). Categorization itself produced data operating with performative force to designate a threat. To put it this way, Matusow's clear unreliability, the far-fetched nature of the claims he made, the obvious issues with the lists he made (remarked upon by the FBI), were disguised by, or possible to pass over, due to the fact that he *had* a list. It was in fact his capacity to *list* rather than his capacity to *observe* which was at the heart of his activities. If the camera captured the images, the listing process which placed individuals in particular categories laid them down and sorted them.

Breaking off

Matusow is dead, for good I hope. (Letter to Billie, 13 January 1954)[24]

By the mid-1950s Matusow wanted to stop informing and remedy some of the damage done. The threads of his life were falling apart. He had broken with the McCarthy camp and had also divorced multiple times.[25] He first admitted to lying whilst witnessing (notably to the Methodist leader Bishop Oxnam, whom he had earlier slurred), whilst still testifying against others. He was then approached by the lawyers Kahn and Cameron, who were working with union leaders and others. Their hope was that admissions of lying from Matusow could be used to reopen cases of jailed 'communists' – amongst them Clinton Jencks. The mine union worker's conviction on the charge of being a CP member whilst signing the Taft Harley Act had been supported by just one witness – the 'bastard' Matusow.[26] The latter's decision to document his activities, obviously likely to produce perjury charges, was agonized (see also Kahn, 1955). He feared the end of celebrity *and* permanent infamy, and at times appeared ready to pull back:

> When I first returned to New York, I planned to write a book ... exposing everything ... Mc C. Counterattack et al ... I was bitter, ... you might say sick ... unimportant ... I was offered a lot of money to do ... and to make a long story short ... I AM NOT WRITING THE BOOK ... Reason number one ... I don't belong in politics. Number two ... Harvey Matusow couldn't die if the book was written ... I would only live in the past ... (Letter to Arvilla, 13 January 1954).[27]

False Witness did get written. Matusow dictated it to Kahn, who transcribed it (Lichtman and Cohen, 2004). Affidavits signed in the run-up to its publication in which Matusow declared he had made false testimony under oath sensationally ended his career as an informer and threatened to blow apart the entire system of professional anti-communist informers (Schrecker, 1998 and Schrecker cited in Lichtman and Cohen, 2004: 162). It did open the way for charges to be laid against him. This wasn't straightforward, since victims of his testimony who had gone down as communists could not be used as defence witnesses, but material involving Roy Cohn could be and was used (Lichtman and Cohen, 2004).[28] Kahn notes that, ironically, this meant that Matusow was never indicted for 'what he had confessed to in *False Witness* ... lying as a government witness ... helping railroad people to jail, for blacklisting and ruining the reputation of scores of citizens' (Kahn, 1987: 253).

In August 1956 Matusow went to prison. He served some time in Lewisburg Federal Penitentiary in Pennsylvania, where the librarian was Wilhelm Reich, known to fellow cons as the 'sex-box' man (Elmer Swink, *IT*, 8 May 1956).[29] He was paroled early and was discharged from supervision on 3 September 1961, typically saving the parole board certificate from the Federal Prison in Allenwood, PA.[30] Six years later he left America. He had tried to evade his earlier life but, suitably enough, was constantly informed upon. His decision to leave was apparently made after a hostile meeting with singer Pete Seeger's wife, who held him partly responsible for her husband's blacklisting. He went to England.

The man 'punching holes in computers'

In 1970 *Oz*, the counter-cultural magazine, launched issue 28, the Schoolkids issue, and went to court in a famous obscenity case that sentenced two of its three founders to prison (Sutherland, 1982).[31] *Oz* was notorious as an organ of 1960s generational rebellion and, according to Charles Shah Murray, one of the kids, the issue arose out of a desire to interrogate 'actual rather than notional' kids (Murray, 2001). *Oz* had a clear sense of the connections between medium, media style, and political intent. It set out to redefine the publishing format. In his account John Sutherland sums it up as 'a polychrome mélange. Pages changed colour, print came from all directions in a vertiginous riot of typography and disorderly layout ... it had no set format. One issue was entirely pictorial – an "image bank" with no words attached' (Sutherland, 1982: 117). In accord with this, in notes for a project entitled 'New Horizons', *Oz* co-founder Richard Neville, noting the importance of sheer presentation to movements at the time, described the alternative society as televisual.[32] Of *Oz* he said that 'in terms of its topography, its use of colours, design and layout, it's ... much more akin, I'd say to a colour TV screen than it is to old fashioned molten lead bits of Grey A point type publications which are really the result of a sort of nineteenth century way of thinking, ... sort of McLuhanistic, tactile and lateral ... '.[33]

All the more surprising, then, that on page 45 of the 'Kids' issue, somewhere after a notorious cartoon showing Rupert Bear doing unimaginable things that shocked English judges, were two starkly printed and formally laid out black and white pages. This was database form 'OHID'. The acronym stood for On Human Individual Dignity (*Oz*, 1970).[34] The form demanded that users fill in boxes giving information on 'information quota allowances', made demands for 'information averages', information 'gains' or 'losses', and 'sale, exchange or involuntary conversation of said information', and set out 'alternative information computation' boxes. In short, it was a spoof. The author of this medium-theoretic contribution was Matusow. In a magazine designed for those whose sensibilities were consciously televisual, he was satirizing computing, databases and ways in which the computational might be used to redefine, capture, or threaten identity.³⁵ If *Oz* was beyond print aesthetics, Matusow was beyond the image and into information.³⁶

How did he get there? That is, how did he come to hold this position, what was he doing in *Oz*? Arriving in England, Matusow became a journalist and broadcaster. He had some mainstream success (e.g. talks to BBC Radio 4, the Third Programme, work with trade press publications), but also became a counter-cultural journalist, contributing to underground and alternative magazines, including *Oz* and more often *IT*. He associated with the London Film Co-Op, became a filmmaker,³⁷ led a band which produced one (very relatively) successful album, the *War between Fats and Thins*,³⁸ and engaged in various acts of quasi-situationalist activism. He also collected taxis.

The arrival in England amounted to a reinvention. Matusow's journalism did reference the US, and even McCarthyism³⁹ – but rarely his own role in it. He wrote as a 'scene' insider and as an émigré, avowedly fascinated by the difference between UK and US politics.⁴⁰ In the *Fats and Thins* he invoked his past in indirect terms, but a more sustained engagement is in the *Stringless YoYo*, a film produced in 1961, which collages HUAC hearings and Nazi rallies, both run backwards to the sound of dance music, and the swirls of a stringless yoyo, which latter are used to produce psychedelic Bridget Riley-style graphical effects. The eponymous yoyo had a bit part in the recantation hearings and revealed the absurdity of the process. At one point, captured on the film, Matusow discusses his invention of the yoyo and is asked by his accusers whether the manufacturer is or has ever been a communist.⁴¹ As a database compilation of a series of archive assets, the film might be an early post-narrative database film. It also indexes Matusow's developing views on computing itself, thereby bridging his earlier and later activities. Sometime later, writing about the anti-computing league, he summed this up when he declared, 'I don't like fascism, don't like bureaucracy, I don't like technology' (cited in Berenyi, 1971).⁴²

Anti-computing: database anxiety and the anti-computing league (computerized collecting)

'Should we ... organize mass demonstrations, march on IBM? I don't think so. If nothing else this is a cool organization. It's the computer that's going to get short-circuited. It's we who communicate. (October 1969)[43]

By the mid- to late 1960s computers were becoming more visible in UK society, and were not only being represented as reclusive instruments for giant corporations, governments, or as Cold War weaponry (Stone and Warner, 1970). Their further expansion was also understood as imminent. A new job category – computer programmer – arose and was widely advertised. New uses for computers in offices and businesses generally were under development. The *Daily Mirror* commented that in the US 'the Mets won the world series and the bulk of paper thrown out of New York streets turned out to be computer punchcards'. The *Mirror* added that this computer 'revolution' was coming England's way (*Daily Mirror*, 17 April 1969).

Matusow's growing dislike of computational technology was already evident in the *Stringless Yoyo*. In the few years following, he wrote anti-computing articles for various outlets. Technology became a recurring theme in his columns on 'Anti-Matter', written for *American in London*. Drafts in the collection[44] include stories on social shaping and technology, the unintended consequences of innovation (e.g. the cotton gin), the computerization of personnel management in the US army as its difficulties with Vietnam escalated (November 1970, column on Vietnam).[45] In fragments of a book draft Matusow makes a case for Vietnam as a 'computer atrocity story'.[46] There are also articles on computerization and political process. In a discussion that seems prescient in the light of the later ballot-box scandals of the Bush/Gore era and Trump's use of social media, Matusow noted dryly that 'now we have a computerized president'.[47]

However, the major focus of Matusow's anti-computing activities concerned the consequences for individuals of the computerization of everyday life, particularly life in the UK, specifically in relation to databases.[48] Around 1968 or earlier, declaring himself 'interested in the abuses of the computer' he founded an anti-computing society. The International Society for the Abolition of Data Processing Machines, or ISADPM, which took off a few years later, and had some reach by the late 1960s, had as its aim 'the destruction of man's over-dependence on the computer' (Matusow, 1968b: 7). In the UK it gained a membership of up to 7,000 people and had some resonance. It helped to make Matusow an unofficial spokesman for anti-computing sentiment in general for a period. The Society was always most active in England, but there were copycat outfits in other countries (e.g. in New Zealand and Germany) and some ISADPM material, including the 'atrocities' book (*Beast of Business*, see later), was translated into German. There were also claimed links with Japan.[49]

The ISADPM logo was a catapult, a 'symbol for battle', man pitted against the giant machine. Matusow recognized that a sling would be more correct, but he didn't think a total war against machines could be won anyway. Of his society he said we 'are not Luddites as such' (see *Daily Mirror*, 16 January 1969: 11).[50] The battle was 'against over dependence'[51] and the unwanted intrusion of the computer into unnecessary areas. It was allowed that the computer might have a 'constructive function' in 'mathematics and the other sciences'. But when 'the uses of the computer involve business or government ... the individual is tyrannized – and ... society makes its stand' (Matusow, 1968b: 22).

Matusow's hostility came in three parts. He raised practical objections to (what he saw as) an over-reliance on computers, and in this context also objected to what he saw as a widespread over-estimation of their powers: 'firms which use them expect too much from them ... They look on the computer as an omnipotent God which can do no wrong and make no errors.' More fundamentally, he feared computer autonomy: 'the computer companies are turning more and more to ultra-sophisticated computers ...which they hope will develop a sense of reasoning ... for me this is too much – computers programming computers' (Matusow, 1968b: 23). Underpinning these objections was a more general hostility to the computational mode. Matusow feared computers would bring a 'pure, clean and sterile world', introducing modes of uniformity and systematization, and that they would undermine the individual, threatening their autonomy, and remaking them. People would be retro-fitted to conform to programming needs rather than the other way around (Matusow, 1968b: 23). If one feared outcome was machinic individuals, the other concern was that people would become the human individual the computer declared them to be. Matusow often returns to the consequences of computer error, concerned that assumption of computational infallibility would entail the presumption of human fault, and also worried that errors made would be increasingly difficult to correct. These concerns are repeatedly tied in to questions of reputation and identity: 'if any information is incorrectly programmed into the machine, no one questions it – there is no recourse and a person's reputation can be seriously hurt by an incorrect computer statement' (Matusow, 1968b: 23).

Matusow's early anti-computing commentary was in journals. In the mid-1960s he looked for a publisher for a book on 'computer atrocities', to be light-hearted but with serious intent. The *Beast of Business* was published in 1968. It declared itself a 'record of [computer] atrocities against the human race ... and a guerrilla warfare manual for striking back'. Matusow compiled and edited 'atrocity' contributions, which were also solicited by the publishers.[52] He also wrote a framing essay declaring computerization an affront to human dignity, reducing freedom, introducing specific un-freedoms, and rendering societies into bureaucratic monoliths unable to escape the ludicrous excesses of their own rule-makers.[53] The *Beast*'s front cover declared:

> The war is intensifying very rapidly. Be ready for its consequences. You have little time before it engulfs you to learn to know that the computer is here and to show you that you know what its dangers are. At a moment's notice you may be involved in a situation in a government office, or with the police, your bank or a post office. In that critical hour you will be alone and defenseless against the statistics and data which the machine pours out. You must be prepared. You cannot tell under what conditions and at what hour you may have to fight the machine. You want to know what is expected of you. You must be ready for any eventuality. (Matusow, 1968a)[54]

Matusow invited those interested in action against the threat of computerization to join his society (see e.g. *Daily Mirror*, 16 January 1969). He was back in an organization, of a kind. This one was in part a media construct, although it was more than that. What it was not was a 'standard' activist organization, nor was it attached to organized politics. Consonant with his affiliation to *Oz*, the society had more to do with the counter-culture than the practices of the Leninist Left(s). In a letter to putative members ('Dear Individual

...'), Matusow addressed the perennial (for revolutionaries) question of What Is to Be Done:

> Should we ... organize mass demonstrations, march on IBM? I don't think so. If nothing else this is a cool organization. It's the computer that's going to get short-circuited. It's we who communicate. (October 1969)[55]

The proposed theatre of disruptive and 'cool' activity extended from the sphere of personal, into the social arena, and into the corporate world. The society backed direct action of various kinds, from the pedantic through the absurd, and on to potentially serious acts of sabotage. Society members were said to add extra holes to punch cards to disrupt automated personal bills and records processes. Other sabotage methods discussed were rubbing clear wax on the space on forms designated 'for official use only'. PHS in the *Times* diary invoked the society's suggestion that census forms should be thwarted by inserting 'not legally competent to answer' wherever possible, rather than obediently contributing to the exercise of data collection. Getting into the spirit of things, PHS also suggested using the OHID forms Matusow had generated for *Oz* (*The Times*, 15 April 1971). Most of the activities discussed were more talked of than undertaken. Sabotage, even that which was reported as already taking place, was largely a rhetorical possibility.

The society's most active and influential phase began when Matusow gave an interview on John Peel's *Nightride* show (1969), discussing his fears about computer databases and the social world. This produced a strong public response[56] and was picked up by other press and TV organizations. Matusow fanned the flames, giving interviews, writing letters to editors (e.g. to *The Times*, 24 August 1970, saying all errors are not human after all ...).[57] Reports followed in national publications, in the main more attracted by the issue than the society itself. *Dataweek* was representative when it commented that the society, 'though quite small', was interesting because of the public response it had provoked.[58]

Press reports on membership numbers of Matusow's organization varied. A column by Chris Ward in the *Mirror* (17 April 1969: 7),[59] headlined 'How to Hit a Computer Where It Hurts', announced that 'more than 5000 computers are now working for – or some would say against – us in Britain and this number is increasing daily'.

Ward noted that human opponents of computers were growing in response. He cited Matusow's prediction that 'very soon ... numbers [of supporters] "will outnumber computers"'.[60] An *Evening Echo* report on 27 August 1970 by Jean Ritchie, headlined 'It's Them Or Us', talked of 'freedom fighting anti computer people' and of an organization with 5,000 subscribing members.[61] Later membership figures quoted rose to 6,000 or 7,000 (although membership databases, particularly those of political organizations, are notoriously massaged, as Matusow himself was abundantly aware).

The society claimed international tendencies. In response to a *Sunday Times* article, Matusow pointed to a New Zealand chapter of ISADPM which declared itself a 'computer control group' (Atticus, *Sunday Times*, 2 September 1969). There was some coverage in international mainstream media (e.g. the *Sunday Herald*, in Australia, noted 'Computer Hate Body Formed' in September 1968, and the *Vancouver Sun* covered the same story in 1969).[62] There was also exposure in left and alternative presses outside the UK, e.g. in *Automatisme* (Boribard, 1970).[63] Managing the interest and the numbers of contacts by hand was an issue. Ward notes that Matusow was aware of the paradoxical nature of this difficulty: 'what I could do with, around here, he said, is a computer' (17 April 1969: 7).[64]

Appeal, reach

The appeal of the anti-computing league and its ideas was diverse. It certainly linked into counter-cultural suspicions of the Machine, as Matusow's work for *IT* and *Oz* attests to. It also attached to a Middle England dislike of both modernization and 'interference' in personal or intimate life. Finally, it attracted some computer professionals; reviews of *Beast of Business* and reports on the society appear in various counter-cultural magazines/zines for data processor professionals. *Real Time* (strapline: 'communication between computer people') included society coverage in Issue 2 and ran an editorial in Issue 3. Matusow himself claimed that:

> there are over 50 IBM employees who are members of the International Society for the Abolition of Data Processing Machines. Exactly what they do to screw up IBM operations, they are not telling, but rest

assured, they are doing their bit, and IBM high up executives have cause to lose some sleep.

IBM apparently did not respond. It did, however, turn down a suggestion that it should back the book; with his usual ethical flexibility Matusow had suggested this on the basis that it was more about the need for 'computer PR' than an entirely hostile enterprise. In a column in a business-to-business magazine *Office Methods and Machines*, he reprised this under the headline 'The Computer Needs a PR Man' (May 1969: 14).[65]

Although the press 'made' the ISADPM, arguably it did so because it recognized that the ISADPM stance resonated with a structure of feeling circulating in the UK at the time. Matusow made much of the response he received after the Peel show, though he kept very few of the letters he received in his archive. However, as an inveterate press clipper of his own work, he did archive discussions appearing in the press of the public's response to his book. In 'Computers Under Fire', an article for an issue of *New Scientist* specifically exploring computers and society, Rex Malik,[66] a 'free-lance writer in computer sciences', suggests that the 'prolific flow of letters to Mr Harvey Matusow's *ISADPM* … reflects the widespread irritation and distrust generated by computers' (Malik, 1969: 292).

Some of the letters are abject: 'Dear Mr Matusow, I would like to help wreck computers. Yours sincerely'. Amongst those Malik cites are letters from people with no engagement with computing, apparently unnerved by their proximity to them; one letter writer announced that '[t]he University of Essex is a stronghold of computers and I dislike them greatly'. (Malik, 1969: 292). Many critiqued the assumption that 'it must be true, it came out of the computer', and Malik also highlighted explicit fears about computer use 'in the organization of people without social control'; writer after writer fears 'we might hand over too much to the machine', as one of the correspondents put it. Malik suggested the letters reflected 'an undercurrent of fear about the computer and its effects among the population at large'. The fears they express include concerns raised by Matusow directly – particularly his concern that 'those who do not fit in well with the assumptions of a system could be mistreated by it'. It seems clear that the unclassifiable Mr Matusow, the man with twelve lives, would himself fit into this category. Malik's summation was that, 'despite the absurd title of his society ... [Matusow] ... tells of a kind of nightmare – where a computer outcasts a man – which cannot be ignored ... '.

Reviews of *Beast of Business* in the mainstream and business press varied. *Banker's Magazine* excoriated the book as having 'all the attractions of a kind of computer Powellism ...',[67] arguing that whatever Matusow's own more existential or radical position, the general public was exercised about computers largely around consumer issues. Kinder reviews also suggested that much of the dissent Matusow had tapped into amounted to concerns around the intrusive nuisance of direct mail. The *Sun*, introducing Matusow as the man 'punching holes in computers', did not question his ideological objections to computers, but reported on practical issues concerning automation and the threat to local post offices. (Richard Last, *Sun*, 3 May 1969: 5).[68] There is little about industrial sabotage, nor the military industrial complex, mentioned in these more popular mainstream accounts. There was a tendency to agree that Matusow's attack 'with humour' (*The Office*, August 1971: 30–34) belied a more serious intent. An *Observer* 'Pendennis' column reported on Matusow's call to 'Worry a Computer, Confuse a Computer, Wreck a Computer', alongside a picture of the man (the suits of the HUAC days disappeared as the young man of the 1950s became the 46-year-old hippie). It noted that, despite his peccadillos, including his penchant for the Jew's harp and for collecting taxis, he was 'more serious than you might think – at least on automation'.[69]

Elsewhere the society's aims were represented in more politicized ways. In the alternative press what was elsewhere described as a polite lament for the increasingly bureaucratic management of everyday life (good bill payers thwarted) produced a more thoroughgoing, if indirect and often surreal, attack on bureaucratic structures and their logics of total or permanent capture. Volume 2 of *Real Time* magazine[70] discussed anti-computing in an editorial and previewed a forthcoming issue on Matusow and the society with the bald question 'are computers any use whatsoever?' Reviewing the *Beast of Business*, it declares, more or less approvingly, that it will explain how '[T]he computer is there to serve you, not to be your master. It's a guerilla warfare manual for striking back at the creeping menace ... From it you can learn how to de-magnetize your cheques, add millions to your computerized bank statement, get ten tons of broken biscuits delivered to people you don't like and generally how to worry, confuse, and wreck a computer.'[71] Volume 3 does not appear in the archive – and may not have materialized.

The ease with which the society fitted itself into all these various organs suggests its founder's continuing ideological flexibility. Its Board included musician Anna Lockwood (then Matusow's wife) and individuals from Wolfe, the publishers of the 'anti-computer book'.[72] Clearly, the society, whatever else it was designed to do, also functioned simply as a marketing tool for the book. Other PR work was done too; the archive contains press shots of Harvey strangling, or being strangled by, tape in front of the reel-to-reel-like cases of a bank of computers.[73]

In finding the anti-computer cause, Matusow had arguably found a new 'gimmick' and was setting out to exploit it. Once again, in a new and much diminished sphere, he became an 'expert' witness and dined out on this activity.[74] He became known as an authority, speaking for the anti-computing 'side' in many debates around computerization, databases, and society. He was eschewed by some for his populism and obvious showmanship. IBM not only didn't back the book but declined to accept him as a spokesman in a debate about computers and society, stating that his involvement would not be at a 'worthwhile level' (*Sunday Times*, 9 January 1972: 32).[75] However, he attended other events where the debate was serious and sustained, including, for instance, a National Council for Civil Liberties forum on civil liberties and the 'Databank Society',[76] and his views were taken seriously by various MPs and politicians, including some involved in discussions around computerization and databases in relation to Kenneth Baker's 1969 Bill on Right of Privacy.

A consistent thread in the writing of Matusow at this period concerns exoneration. He repeatedly expresses a fear and a belief that the computerized collection of data on human individuals is a one-way street; once in, never out. The difficulty of deletion for those named on a list was also of course integral to the blacklisting operations of the McCarthy era (as it was, as Miller had noted, to the witch-hunts of Salem). Matusow saw the same dynamic threatening to return via computerization – where it would also become a general condition, and one that might, by virtue of its superior collecting capacity, be far more absolute. He feared that the computer industry might 'like Moby Dick' swallow us whole.[77]

The end of 'anti-computing' …

If it happened, we didn't notice. Or perhaps we are in the belly of the beast. This early anti-computing moment, with its focus on the consequences of data capture and storage for personal freedom had faded by the mid-1970s. Computers became ubiquitous in offices, working with them was far more routine, and gaming arrived, in the UK at least, in the early 1980s (see Haddon, 1988), offering a different mode of interaction. At around the same time the personal computer was introduced, undermining for the moment visions of the centralized database society and explicitly challenging the IBM model of computation. Paradoxically, this challenge came from Apple, responsible for one of the most successful closed computing systems ever launched. In another paradox, one of the legacies of the underground press with which Matusow was involved was the rise of listings magazines (e.g. *Time Out*). As for the *Beast*, it went out of print and is now available only second hand through various online outlets. As one seller wryly comments, this in itself 'proved we lost'.[78]

Matusow himself left England. In his later years he spent time in austere low-tech communities in the US. Kahn tells us he was at times without even electricity. He did later revise his anti-computing stance, surfacing rather early on the web as Job Matusow, via a website called CockyBOO, which appears to be an attempt at a cooperative biography. Matusow died during its construction and the page became an early example of digital memorialization. The tributes there are often rather affectionate. They are for Job (Harvey) Matusow and are largely read by the odd few curious and uncertain onlookers, circulating around a now dead but still unfixed identity – 'are you the same Matusow?' one asks.

Matusow could be dismissed as a man entirely without consistency whose anti-computing activities were only a peculiar footnote in a life whose activities had been significant on a far larger stage. What has been argued here is that Matusow's hostility to the computational database and its operations, his concern with the relationship between the computer record and its operationalization, his focus on how what is collected is put to work, all find echoes in his earlier activities. So, if understanding 'the truth' of his motivations is impossible, not least because these motivations shift and dissolve, then we can note that his ambiguous relation to media, to the spectacle, but above all to the *record*, is remarkably consistent in its ambiguity. It is characteristic that more than once he declared his dislike for technology, and computer technology, in magazines dedicated to their promulgation.

Conclusion 1: culpability

To explore the onion layers of Harvey Matusow's history is to become aware that his most constant trait was a constant tendency towards constant revisionism. This appeared to be enabled by a sense that his version of events could have credibility or force, even if demonstrably not true – something that now looks horribly familiar, particularly in relation to political leaders. Decades later, in an interview and report by David Madison (1998) from filmvault.com, Matusow represents himself as a peculiar kind of hero, rewriting his informing activities one more time to suggest he was 'toying with Congress and the country's salivating media … telling a string of outrageous lies that McCarthy and the others never questioned'. This is at one with other moments in which he appears to be revising his recantation story.[79] Perhaps, after terminal exposure, and after his activities to expose others, and finally in the light of this late and disgraceful attempt at exculpation, he deserves oblivion. What seems more productive, however, is to observe the threads and events of his life, to understand *that* this engagement with the monstrously miscommunicating systems of the McCarthy era made him somebody opposed to database power, who understood databases as the powerful means through which people might be engulfed by large corporations, organizations, or states.

Conclusion 2: computer history

Matusow's story contributes to histories of the counter-cultural engagement with computer technologies. Antipathy to the Machine, the latter nomenclature designating both a political system (nascent global capitalism and the post-war world) and the hard, technocratic rationality that characterized it, which was part of the broader counter-cultural formation, might be taken to include antagonism to the specifically computational. A series of histories of the personal computing industry in Silicon Valley and its virtual suburbs have demonstrated that this history of alignment is more complex and certainly more partial than is often suggested. Such works have explored the diverse connections between the technology innovators and the counter-cultural movements of the 1960s and early 1970s (Levy, 1984, Cringley, 1996, Turner, 2006), and there is also the evidence of this to be found in *Whole Earth Review* (and perhaps in publications such as Ted Nelson's *Computer Liberation*). Fred Turner's work on Stuart Brand (the *Whole Earth* creator), for instance, traces out how, albeit in relatively small circles, a shift 'into' computer technology was made.

Turner accounts for what turned the US West Coast counter-culture towards digital technologies (what turned them on, technically perhaps), what made them re-evaluate technologies that had been developed as part of military agendas (the early Cold War in particular), and what led to the sustained engagement between the counter-culture and the Valley; this is the shift from the *Whole Earth* to *Wired* via Stuart Brand. The Matusow case suggests an alternative and supplementary trajectory exposing a form of engagement with emerging computer technologies in the late 1960s/early 1970s that is also essentially counter-cultural, but that did not follow a trajectory from hostility (to computer technology as military technology or the technology of the Man) to avidity. This is not necessarily because Matusow lacked the close technical relationship to computer technology that the hackers had, although he did lack that, since his engagement in film and the visual and sonic arts was analogue. Rather, what seem significant are his experiences of databases in operation, and in operation in specifically political contexts. This perhaps led both to his hostility to computers and to a more material series of objections to computerized culture than those arising simply from a principled objection to the (abstract) Machine, based on a general critique of technocratic culture. By the 1960s he was already an expert, not in computer circuitry but in the potentially devastating effects of particular forms of instantiated database operation.

The historical context of the mid- to late 20th-century information society debates was the Cold War, as it developed in the 1940s and the 1950s, and its permanent security imperatives (see e.g. Edwards, 1996). Turner's account considers how, in certain areas, this early conditioning (the fashioning of technology as weaponry) was overcome (tactically perhaps) as the technologically savvy turned (retuned) technology to their own (counter-cultural/later commercial) ends. In this way a counter-culture became a *Wired* culture. Matusow, who experienced the Cold War not through close-up engagement with its technologies but through close-up engagement with its political operations, the anti-communist America of the late 1940s and 1950s,[80] comes to take a different view. Specifically, he was interested in what might be termed, drawing on Latour, the obduracy of the categorizing operations that the computerized sort could undertake. Computerization is for him often an atrocity (even though he is also not entirely against it), and it is clear in fact, reading his papers, that his sense of *why* it is an atrocity comes from his understanding of *how* it might be used as a social weapon. *After all, he has himself been caught up in the use of databases, and the use of data entry, storage and retrieval, to these ends.*

Conclusion 3: media archaeological witnessing?

John Durham Peters, cited above, explores witnessing as a common 'but rarely examined' term in both professional and academic analysis of media events. For Durham Peters, witnessing raises questions of 'truth and experience, presence and absence, death and pain, seeing and saying, and the trustworthiness of perception – in short fundamental questions of communication'. He also views the witness as taking on responsibility so that '[t]o witness an event is to be in some way responsible to it' (Durham Peters, 2001: 707–708). Durham Peters moreover articulates his discussion in medium terms, talking of witnesses as a 'fallible transmission and storage mechanism' (Durham Peters, 2001: 710). Carrie Rentschler, following Durham Peters, develops his sense that the witness is also a medium, and the medium can constitute a form of storage space for witnessed material (Rentschler, 2004).

For Durham Peters, 'in the moral sense: to witness means to be on the right side' (Durham Peters, 2001: 714), because witnessing is a 'mode of communication'. Matusow, in the witness stand, is in this sense not witnessing at all. The storage mechanism named Harvey Matusow was thus *falsely* understood as witnessing. This mechanism is both the man and the medium; the camera apparatus and the apparatus of the hearings. Testimony and technology are related. Indeed, as Durham Peters notes, drawing on Shapin, it was modern science that overcame 'the low repute of testimony'. Writing of modest witnessing Donna Haraway makes a similar argument (Haraway, 1997). Both note that it is because of science and its instruments that testimony came to be trusted: 'thing-like, and hence credible' in its indifference to human interests.

The camera and the microphone both inherit this tradition of objectivity as passivity, says Durham Peters. They produce a form of assumed credibility and reliability that finds its way into court – that is, '[l]egal rules prefer a mechanical witness'. Durham Peters goes on to argue that '[T]he ideal human would behave like a thing: a mere tablet of performing', a 'dumb witness' who has no motive to 'lend comfort' (Durham Peters, 2001: 715–716). In the end, it might be this, the threat that computing could turn human society into something automated so that it is run by a system with no motive to 'lend comfort', that the later Matusow, informed by his own experiences as a storage mechanism, the witness with the camera, the speaker with the microphone at the hearing stand, came to hate.

These final observations on history, the database, and the witness connect. They constitute three sites in which this media archaeology produces a reappraisal, a new possible route to understanding. To stress, these are not replacement 'causal chains' (say for some entirely other reading of the counter-culture and its technological orientation), nor is this an 'escape into history' (Huhtamo and Parikka, 2011: 4). What I am driving at is the restitution of under-emphasized continuities and connections, and a rethinking of computing and its ongoing and historical insertion into a mediated world. In his account of what he always refused to term media archaeology, Zielinski set out to understand (far) earlier forms of media and media work. He did so through what he describes as works of 'praise and commendation' (Huhtamo and Parikka, 2001: 11). I have no commendation, but would not wish to condemn, since it doesn't seem productive. The intention has been, rather, to work through a life history, viewing it as a matter of collection, storage, and retrieval, not to reduce it to its essentials, nor let it stand for an entire moment, but to understand it more fully.

Notes

1. In 2013, Edward Snowden, a Central Intelligence Agency employee and sub-contractor leaked classified information from the National Security Agency to *The Guardian* and *The Washington Post*. The thousands of leaked documents revealed the extent of the mass surveillance programme in the US and produced fierce debate around freedom, digital surveillance, and privacy.
2. A column in the *Daily Mirror*, 17 April 1969 by Chris Ward notes that there are 5,000 computers in the UK and around the same number of members joining Matusow's society [MC Box 21].
3. From Matusow and Malik draft [MC Box 21/2c], and in *The Beast of Business, a Record of Computer Atrocities* (Matusow, 1968a). Dedicated to his mother.
4. Congressman Gallagher. Comments from papers on a 'workshop on the data bank society' [MC Box 21/5].
5. *New Scientist* (17 August 1969) includes Wedgwood Benn on 'maintaining human supremacy' (274–275) and Malik on 'Computers Under Fire' (292) reporting on prolific letters to HM and declaring this reflects 'widespread irritation and disgust generated by computers' [MC Box 21].
6. [MC Box 4:1].
7. Will and testament (1969) [MC Box 4:1].
8. 'Like sniffing a mildewed Facebook page', as one of my colleagues put it.
9. Comments in his film, the *Stringless YoYo*, do not necessarily reassure. There Matusow said the truth is not 'a narrow line' but, rather, has 'a certain direction' (1961) (South East Film and Video Archive – SEFVA).
10. [MC Box 3:1].
11. [MC Box 3:1].
12. Footage at SEFVA. Date is given as late 1950s (Reel 1). Goddard's question to Matusow is 'How many lies do you estimate you have told?'
13. A copy of Ernst and Loth's *Report on the American Communist* (1952) annotated by Matusow is held in the collection. It claims 'on average the typical Communist is a party member for possibly two or three years' (Ernst and Loth, 1952: 14).

14 Matusow was expelled from the Tompkins Square (New York City) section but retained some anonymity outside the East coast and continued to work as an undercover informer there for some time.
15 The 1951 film *I Was a Communist for the FBI*, directed by Gordon Douglas, drew on Cvetic's experiences.
16 Letter [MC Box 7].
17 'Thou shalt not bear false witness against thy neighbour'.
18 He also became a professional speaker. See e.g. a flyer for 'Harvey Matusow "it can happen here"', a lecture at the Libertyville Community Sunday Evening Club (5 October 1952), which claims 'Matusow's lectures answer the question ... "how can our young boys and girls become Reds?"' [MC Box 6/3, 6/4].
19 [MC Box xi].
20 [MC Box 6/2].
21 [MC Box X (h)].
22 A 'Clark H. Gatts Presents' flyer names Matusow a 'special investigator with the Ohio State Committee on Un-American Activities' [MC Box 6/4 (h)].
23 Schrecker (2002) argues that without the participation of the private sector McCarthyism would not have affected the rank-and-file members of the communist movement so greatly, nor have so effectively stifled political dissent.
24 Letter to Arvilla. 'Harvey Matusow couldn't die if the book was written' [MC Box 6/4].
25 'Matusow fails to talk wife out of divorce, quits Reno' (*Times Herald*, 27 September 1953). This was their second divorce in five weeks. 'Rich wife tosses out ex-Red as he asks her to drop suit' (*Reno News*, 25 September 1953) [MC Box 6/4].
26 For a full account, including discussion of the Justice Department's bid to seize the book, see Kahn's *The Matusow Affair* (1987).
27 [MC Box 6/4].
28 Matusow on Cohn: 'It's not secret I don't like him. It was his twisted testimony that sent me to prison for four years ...' ('Who Fought for Bumpy?', 'Anti-Matter' column, *The American Abroad*, 14 December 1968 [MC Box 37]. See also Lichtman and Cohen (2004).
29 Matusow was engaged in prison issues for the remainder of his days. *Oz* (December 1969) describes him as a man who turned people on to the plight of (hippie) prisoners held in overseas jails for drugs offences first in Turkey and in the Lebanon. See also Matusow, 'Sexual Brutality in Prison', *Forum* (1979), 2(12): 26–30, London: Forum Press [MC Box 37/3].
30 [MC Box 4/3].

31 The third *Oz* founder, Felix Dennis, who escaped prison, later founded a computer magazine publishing house. *Oz* 28 is part of the Matusow collection.
32 Information packs for the defence of *Oz* declare it to be 'constitutionally incapable of facing a solemn fun free future … cutting cane beneath some Spartan banner of liberation'. A flyer for the Independence Day Carnival, Sunday 4 July, Hyde Park, in the run-up to the *Oz* trial tells the same story (see also *Oz*, December 1969 issue). Much of Matusow's later writing tunes with *Oz*'s attempt to slip between consumer capital and the rigidities of orthodox left politics [MC Box 37/3].
33 See also Matusow on Neville and medium form in a draft review of Richard Neville's 1970s *Playpower*, submitted to *IT* magazine. 'McLuhan is dead – long live Neville, who has found a new way to use words in a book – he writes with his eyes' [MC Box 37, MC Box 49/6].
34 A copy of the School Kids issue of *Oz* (issue 28) is in the collection – as are separate copies of the Human Dignity forms [MC Box 37].
35 A draft script was prepared for a 'New Horizons' or 'alternative society' documentary project to involve Matusow and Neville. Neville's comments on the televisual are included here [MC Box 49/6]. Others were also involved in questioning media and mediums and Matusow was well aware of them. The collection includes the first issue of *Cinemantics* (1970), striking for artist Malcolm Le Grice's (1970) article on television, effectively a pre-emptive demand for a (utopian version of) an internet for the people [MC Box 49/13].
36 Predictions are interesting. See e.g. 'Mr Matusow looks forward to the time when one will be able to buy an LP of a film which can be played through a television set. The Japanese have already developed the technique but it is not economic for home use at the present …' (Cameron Hill, Interview with HM, *New Zealand Listener*, 30 June 1967) [MC 37/2].
37 *Pot*, one of Matusow's later films, attempted to discover if it was possible to make a film on acid [MC 37/2].
38 Songs available at http://blog.wfmu.org/freeform/2006/07/harvey_matusow_.html.
39 See e.g. Matusow's 'McCarthyism: Could It Happen Again?' for Daniel Snowman on the BBC Third Programme (transmission 12 November 1969, 18:57hrs).
40 Matusow summed this up in a column on the Grosvenor Square riot, which he said had tea breaks.
41 *IT*, May 8th, no year visible [MC Box 37/2].
42 Clipping from *The Office Overseas* [MC Box 21].

43 Letter to subscribers by Harvey Matusow, October 1969 [MC Box 21/1].
44 [MC Box 21/2c].
45 Supposedly the US government was offering cash rewards for dealing with Vietnam deserters, computerizing the program for dealing with them [MC Box 21/2c]. 'Anti-Matter' columns are in *The American Abroad* (December 1970). Those in the collection include a report on the improbable Spreaders campaign (feeding LSD to grannies), LSD and the US navy (fears of ships being mistaken for carpets of flowers), and the burning of the Esso refinery at Humble (December 1970, ms. for 'Anti-Matter') [MC Box 21/2c].
46 Fragments of a book draft, p. 70 [MC Box 21/2c].
47 'New American' column in *The American Abroad*, 16 November 1969 [MC Box 21/2b].
48 The National Computer Centre was established in July 1966 in Manchester by the Labour government.
49 Letter from Matusow to Kuwajima, April 1970 [MC Box 21/2c]; correspondence from Matusow to Atticus notes a New Zealand 'computer control group' [MC Box 21/1].
50 [MC Box 21/1].
51 [MC Box 21/2a].
52 [MC Box 49/4 (h)].
53 Matusow sought sponsorship for a book on similar themes. E.g. a letter between Matusow and Peter Van Lindonk, apparently of IBM Holland, suggested, 'I'm sure some computer companies would like the idea of sponsoring this book' (Matusow, 6 August 1968). The project is referred to as 'The "anti-computer book"' [MC Box 21 (h)].
54 [MC Box 49/4].
55 [MC Box 21/1].
56 There were 400 letters of support in the *New Scientist*. Pendennis, *Observer* review, 13 April 1969: 38–39 [MC 21/2a] mentions many letters, and the response to Matusow's appearance on John Peel's show in 1969 was vigorous.
57 [MC Box 21/1].
58 *Dataweek*, Comment, 26 February 1969 [MC Box 21].
59 [MC Box 21/2a].
60 According to the Computer History Museum Timeline there were, there were 5,200 computers installed in Britain (www.computerhistory.org/timeline/computers/).
61 [MC Box 21/1].
62 [MC Box 21/3].
63 [MC Box 21/22a].

64 [MC Box 21/2a].
65 [MC Box 21/2c].
66 Rex Malik's article on databanks and the government invokes the Triple Revolution report and 'Cybernation: The Silent Conquest' as failed prophesies (see Chapter 4) [MC Box 21/2c].
67 PES (column) 'The Bankers Bookshelf', *Bankers Magazine* [MC Box 21/1].
68 [MC 21/2a].
69 Pendennis, *Observer* review, 13 April 1969: 38–39 [MC 21/2a].
70 These were micro-publishing outlets; the address of *Real Time* was Hargrave Road, London N16 – now, and probably then, residential.
71 [MC Box 21/2a].
72 Letter from Wolfe Publishing [MC Box 21/2b].
73 [MC Box 21/2c].
74 Matusow kept a letter inviting him to speak after dinner from City of London Round Table – the writer talks of his 'non professional luddite soul' [MC Box 21/3]. Another letter invites him to lunch at the London School of Economics (8 August 1968) [MC box 21/2b].
75 [MC Box 21].
76 Leaflet [MC Box 21/5].
77 From drafted work with Malik [MC Box 21/2c]; A 'workshop on the databank society' [MC Box 21/5].
78 A copy of the book with a dedication from Matusow is to be found in the archive.
79 Lichtman and Cohen (2004) also cover this.
80 As Matusow notes, the 'McCarthy era' began 'before' McCarthy, perhaps in 1947 with the Alger Hiss case [MC Box 37/1].

Bibliography

Apple Computer. 1984. Advertisement for the first Macintosh, directed by Ridley Scott. Quicktime version viewable online at www.apple-history.com/quickgallery.html.
Berenyi, Peter. 1971. 'The Office Overseas "The abomination of computerization"', *The Office* (August): 30–34.
Boribard, P. 1970. *Automatisme*: Tome XV, 9 September (no pagination, fragment).
Cameron, Angus. 1987. 'Introduction', in Albert E. Kahn, *Matusow Affair, Memoir of a National Scandal*, New York: Moyer Bell: xv–xx.
Ceruzzi, Paul E. 2003. *A History of Modern Computing*. London: MIT Press.
Chun, Wendy. 2008. 'Enduring Ephemeral', *Critical Inquiry*, 35: 148–171.

Cringley, Robert X. 1996. *Accidental Empires, How the Boys of Silicon Valley Made Their Millions*. London: Penguin.
Dean, Jodi. 2001. 'Publicity's Secret', *Political Theory*, 29(5): 624–650.
Durham Peters, John. 2001. 'Witnessing', *Media, Culture and Society*, 23: 707–723.
Edwards, Paul N. 1996. *The Closed World: Computers and the Politics of Discourse in Cold War America*, Cambridge, MA: MIT Press.
Elsaesser, Thomas. 2004. 'The New Film History as Media Archaeology', *Cinémas*, 14(2–3): 75–117.
Ernst, Morris L. and David Loth. 1952. *Report on the American Communist*, New York: Henry Holt. [Matusow annotated copy held in Matusow Collection, Special Collections, University of Sussex.]
Fried, Albert. 1997. *McCarthyism, The Great American Red Scare: A Documentary History*. Oxford: Oxford University Press.
Godard, Jean-Luc. Filmed interview with Harvey Matusow. [SEFVA].
Haddon, Leslie. 1988. 'The Home Computer: The Making of a Consumer Electronic', *Science as Culture*, 2: 7–51.
Haraway, Donna and Thyrza Goodeve. 1997. *Modest_Witness@Second_Millennium. FemaleMan©_Meets_OncoMouse™: Feminism and Technoscience*. London: Routledge.
Huhtamo, Erkki and Jussi Parikka. 2011. 'Introduction', in *An Archeology of Media Archeology*. Berkeley: University of California Press, 1–20.
Kahn, Albert E. 1950. *High Treason: The Plot Against the People*. New York: Lear.
Kahn, Albert E. 1955. 'The Story Behind This Book', preface in Harvey Matusow, *False Witness*. New York: Cameron and Kahn, 7–17.
Kahn, Albert E. 1987. *The Matusow Affair, Memoir of a National Scandal*, New York: Moyer Bell.
Kittler, Friedrich. 1997. *Literature, Media, Information Systems*. London: Routledge.
Krutnik, Frank. 2007. *'Un-American' Hollywood: Politics and Film in the Blacklist Era*. New Jersey: Rutgers University Press.
Lattimore, Owen. 1950. *Ordeal by Slander*. New York: Bantam Books.
Le Grice, Malcolm. 1970. 'Outline for a Theory of the Development of Television', *Cinematics*, 1: 215–218.
Levy, Stephen. 1984. *Hackers*. London: Penguin.
Lichtman, Robert M. and Ronald D. Cohen. 2004. *Deadly Farce: Harvey Matusow and the Informer System in the McCarthy Era*. Urbana: University of Illinois Press.
Locke, John. 1975. *An Essay Concerning Human Understanding*. Oxford: Clarendon Press.
Madison, David. 1998. *Interview*, FilmVault.com, filmvault.com/slc/stringlessyoyothe1.html.
Malik, Rex. 1969. 'Computers Under Fire', *New Scientist*, 7 August: 292.
Matusow, Harvey. 1955. *False Witness*. New York: Cameron and Kahn.

Matusow, Harvey. 1961. *The Stringless Yoyo*, film [held at SEFVA]. SxMs8/9/1/1/2. 16mm and VHS cassette SxMs8/9/1/1/3.
Matusow, Harvey (ed.). 1968a. *The Beast of Business, A Record of Computer Atrocities*. London: Wolfe Publishing.
Matusow, Harvey. 1968b. 'Know Your Enemy', Introduction in *The Beast of Business, A Record of Computer Atrocities*. London: Wolfe Publishing.
Matusow, Harvey. 1969. *McCarthyism: Could it Happen Again? With Daniel Snowman*. BBC Third Programme, 12 November. [Partial transcript]
Matusow, Harvey. 1970. 'On Human Dignity', *Oz*, 28: 'The School Kids Issue'. [Originals or earlier copies stored in the Matusow Collection, Special Collections, University of Sussex.]
Miller, Arthur. 2000. '"Are You Now or Were You Ever?"', *The Guardian*, 17 June, www.theguardian.com/books/2000/jun/17/history.politics.
Murray, Charles Shah. 2001. 'I Was an Oz Schoolkid', *The Guardian*, 2 August, www.theguardian.com/media/2001/aug/02/pressandpublishing.g2.
National Council for Civil Liberties. Undated flyer for Open Workshop on Data Banks Privacy and the Public. [MC Box 21/5].
Navasky, Victor. 1980. *Naming Names*. New York: Viking Press.
New Scientist. 1969. 'Wedgwood Benn Exposes the Myth of Technological Tyranny', *New Scientist* cover, 17 August. [Matusow Collection, Special Collections, University of Sussex.]
Oz. 1969. Issue 25. Hippie Atrocities (December).
Oz. 1970. Issue 28. The Schoolkids issue (May).
PHS/Times Diary. 1971. 'Wax and Wane', *The Times*, April [Matusow Collection, Special Collections, University of Sussex.]
Rentschler, Carrie. 2004. 'Witnessing: US Citizenship and the Vicarious Experience of Suffering', *Media, Culture and Society*, 26(2): 296–304.
Rolling Stone. 1972. Interview with Harvey Matusow, 17 August. [Matusow Collection, Special Collections, University of Sussex.]
Schrecker, Ellen. 1998. *Many Are the Crimes: McCarthyism in America*. New York: Little, Brown.
Schrecker, Ellen. 2002. *The Age of McCarthyism: A Brief History with Documents* (2nd edn). London: Palgrave Macmillan.
Stone, Michael and Malcolm Warner. 1970. *The Data Bank Society: Organizations, Computers and Social Freedom*. London: George Allen Unwin.
Sutherland, John. 1982. *Offensive Literature*. London: Junction Books.
Turner, Fred. 2006. *From Counter-culture to Cyberculture: Stuart Brand, the Whole Earth Network, and the Rise of Digital Utopianism*. Chicago: University of Chicago Press.
Weiss, Eric. 1998. 'Review of Paul N. Edwards, The Closed World: Computers and the Politics of Discourse, in *Cold War America, Inside Technology Series*, Cambridge, MA: MIT Press', *Minds and Machines*, 8(3): 463–468.
Zinn, Howard. 1980. *A People's History of the United States*. New York: Harper & Row.

Chapter 4

No special pleading: Arendt, automation, and the cybercultural revolution

Is it possible to attend a conference fifty years after it has ended? The attempt is made in this chapter which explores a mid- to late 20th-century debate around the leisure society and the end of work. The focus is on Hannah Arendt's intervention in a paper recapitulating many arguments developed in her major work *The Human Condition* (1998 [1958]) but laying them out to an interested audience with their own positions. An exploration of the stakes of the early cybernation debates opens up questions about the end of work that find new salience today. Exploring Arendt's work in these contexts sheds fresh light on her analysis of 'a labour society approaching its end' (Lenz, 2005: 135). It is thus also to reassess how Arendt's thinking on technology travels.

* * *

> Cybernation at last forces us to answer the historic questions: 'What is man's role when he is not dependent on his own activities for the material basis of life'? (The Ad Hoc Committee on the Triple Revolution [Agger et al.], 1964).
>
> Man cannot be free if he does not know he is subject to necessity, because his freedom is always won in his never wholly successful attempts to liberate himself from necessity. (Arendt, 1998: 121).

On a hot two days in June 1964[1] an unlikely group, including computer scientists, engineers, philosophers, political scientists, a feminist, civil rights activists, theologians, government workers, and administrators, Labour leaders, entrepreneurs, and a Left hero once accused of spying, assembled at the Hotel Madison in New York

City for the First Annual Conference on the Cybercultural Revolution – Cybernetics and Automation. Cost of attendance, paid in advance, $7.50. This gave entry to two events: the Cybercultural Revolution was held back to back with the Third Annual Conference of the Congress of Scientists on Survival.[2] The conference was convened by Alice Mary Hilton, a programmer and social activist, and engaged with the ideas of the Ad Hoc Committee on Triple Revolution, a coalition of the expert and the interested who saw in computerization, and particularly in cybernation – defined as the convergence of automation and computerization (Michael, 1962) – the possibility of fundamental and radical social change. The Triple Revolution's manifesto identified 'three revolutions underway in the world ... the cybernation revolution of increasing automation; the weaponry revolution of mutually assured destruction; and the human rights revolution' (Boggs, 1963). Peace was the greatest prize, civil rights the most pressing as a political demand, but the *means* identified to bring about change was cybernation (Agger et al., 1964).

Speakers and panellists at the Cybercultural Revolution conference explored the displacement of human labour, managing the transition from human to machine labour, and the future prospects for the society of idleness or leisure to be inaugurated by computers. They set out to define and explore the technologies with which they were dealing, to consider the Evolving (present) and Future Society, and to explore global issues. They were speaking into a world changing very rapidly; key contexts were the Cold War, the bomb, President Johnson and the 'Great Society' programme, the first formal successes for the civil rights movements and the coming long, hot summers of the early to mid-1960s urban riots.[3] These were also the years when mass consumption provided many in the US with a well-equipped home (Cowan, 1985), and when the first stirrings of feminism's discontents with what recompense automated domesticity offered women were felt (Friedan, 1963).[4] Amongst these rapid shifts there was also the developing legacy of first-wave cybernetics, the early expansion of computer machines into business and their continually shrinking footprint (in 1964 IBM launched its first non-monolith computer), and the acceleration of automation through the coupling of mechanized processes with computational technologies or 'computer machines' that enabled cybernetic systemization of production processes.

Amongst contributors to the conference was the philosopher Hannah Arendt, whose typed conference paper 'On The Human Condition' is to be found in the Library of Congress along with a letter from Hilton chasing her for her copy. The paper also appears in *Evolving Society*, a conference proceedings publication edited by Hilton (Hilton, 1966) which included the conference papers, panel discussions, chairman's summations, and many contributions from the floor. Arendt's intervention was much anticipated by many attendees. Her presence was felt in other ways too; her thinking influenced Hilton's organization of themes as well as the latter's own work (Hilton, 1964, and my personal interview). It was a key framework through which the conference explored the potential risks and benefits (and/or the good and evil aspects) of the fully cybernated future that many confidently anticipated would rapidly come about.

This chapter explores Arendt's consideration of cyberculture as she explored it in dialogue with these other voices, during the time of the 'cybernation scare' (Ganz, 1966), and asks how it reframes themes found in *The Human Condition*, published in 1958. It draws on Arendt's paper,[5] on *The Evolving Society* and the *Triple Revolution* report. A key additional resource is an extended interview with Alice Mary Hilton which I conducted in New York in 2010, a few years before her death (Hilton, 2010). Hilton was active around groups interested in social justice and peace and was an inveterate debater, described by one admiring reviewer as a 'Socratic Gadfly' (Riepe, 1967), and in a newspaper profile named as the 'Velvet Voice of Automation' (Sheehy, 1964). She was avowedly half in love with Arendt, fascinated by the latter's intellectual glamour, and engaged passionately with her ideas (Hilton, 2010).

Arendt's brief paper to the Cybercultural conference, 'On The Human Condition', might be dismissed as simply an out-take delivered in response to Hilton's urgent solicitations. The context of the paper's delivery, however, as well as the specific form the arguments took, makes it more than that. Arguments around the changing relations between labour, work, action, and the political life set out in print in the rarefied grounds of *The Human Condition* play differently in a live discussion between the peculiarly ('incurably') informed conference goers, who not only spoke for a series of different interests but often embodied them. Elisabeth Young-Bruehl's account of Arendt is of a 'practical minded person' who wanted 'thoughts and words adequate to the new world ... [who wanted to] dissolve clichés, reject thoughtlessly received ideas ... hackneyed analysis, [to] expose lies and doubletalk' (Young-Bruehl, 2006: 11). Arendt herself defined politics as the 'organization or constitution of the power people have when they come together as talking and acting beings' (Young-Bruehl, 2006: 84). The conference was a place where her thinking on what constituted the political came into contact with the messy 'real world' which did not necessarily respect divisions important in her work.

Arendt's presence at the conference also directs attention to her consideration of the specifically technological, which is somewhat neglected in critical assessments of her writing which focus on her thinking on totalitarianism (albeit in its relationship to technocratic rationality); Arendt's work is surprisingly rarely explored for what it says about technology 'itself'.

Finally, there is a way in which 'On The Human Condition', responding to prospects that seemed imminent during the short period of the cybernation scare, operates on a slightly different time scale than *The Human Condition*. We might say it responds to a subtly different conjuncture, to invoke Hall's term (Hall, 1990, Gilbert, 2019). The import of Arendt's paper is that cybernation may mark the surpassing of the victory of labour over work, and the transmogrification of the latter into a mode of job holding coupled with consumption, which is dealt with at length in *The Human Condition*. What was explored as an extreme imbalance in *The Human Condition* (Arendt, 1998, Lenz, 2005) becomes in the conference paper a terminal pathology.

Cybernation and the automation 'scare'

There was, almost from the beginning, an awareness that computers, on the one hand, and cybernetics on the other, would have major social consequences. Cybernetics was – as a systems theory – a general theory, one claimed to have wide applicability for social processes as well as for mathematical and biological ones. The Macy conferences, one of the major early sites for exploring cybernetics, included alongside information theorists and mathematicians, social scientists, anthropologists, and others (Heims, 1991). Norbert Wiener, whose 1940s work on cybernetics helped found the field, understood not only the potential use of cybernetics to model human systems but also the difficulties and problems that might arise in the attempt (see Wiener, 1961, and Wiener cited in Hilton, 1966b). Amongst his fears was mass upheaval as a consequence of automation's effect on jobs. Wiener thought decisions around the organization of a cybernetic society were urgent. His comment 'the hour is very late, and the choice of good and evil knocks at our door' (Wiener, 1989) was much invoked at the conference; many felt the clock had ticked down and the time of decision was at hand (Hilton, 1966: 144, 385). They weren't alone; by the early 1960s the impacts of cybernetic automation were being felt in the US and the likely acceleration of these impacts was being actively explored. Of particular concern was an expected transformation of work of many kinds through the combination of computers and 'the automated self-regulating machine' (Winthrop, 1966a: 113).

The term cybernation was coined by the political scientist Donald N. Michael, in *The Silent Conquest* (1962), a report for the Centre for the Study of Democratic Institutions. It was a designation intended 'to refer to both automation and computers', being invented for 'convenience' and to avoid the awkwardness of repetition (Michael, 1962: 6). Michael believed cybernation would produce 'a profound difference in kind'. There was, he said, 'every reason to be concerned with the implications of thinking machines', whose capabilities and potentialities were 'unlimited', and whose advent had 'extra-ordinary implications for the emancipation and enslavement of mankind' (Michael, 1962: 9). His report begins 'with the advantages of cybernation' (Michael, 1962: 10), concluding that it is needed for the survival of a democratic system, but goes on to consider problems likely to arise – most immediately mass unemployment, suffered unequally so that dominated groups would bear the brunt of the end of work. Michael postulates the creation of four classes, one comprising the entirely unemployed and one with no more leisure than before – which 'in the case of professionals means very few hours of leisure indeed' (Michael, 1962: 29). This is followed by a discussion of publics, of public opinion, and of the threat to the individual – who may be 'completely swallowed up in statistics'. The *Silent Conquest* is fearful for the future partly because the trajectory embarked upon appears unstoppable; Michael argues that there can be no 'moratorium on cybernation'. In a discussion of life 'after the take-over' he wonders gloomily how members of this new society will occupy themselves – 'even with a college education, what will they do all their long lives, day after day, four day week after four day week, vacation after vacation, in a more and more crowded world ... What will they believe in and aspire to as they work their shorter hours, and ... pursue their self-fulfilling activities whatever they may be?' (Michael, 1962: 45). *The Silent Conquest* is a peculiar document read from this distance, at once detailed and speculative, prescriptive and bewildered.

Others were considering similar questions. In England Sir Leon Bagrit gave the BBC Reith Lecture series for 1964 on *The Age of Automation* (Bagrit, 1965, Huhtamo, 1999: 1). In the US the *Silent Conquest* gained some media traction; the *New York Times* (29 January 1962), for instance, writing it up under the headline 'Automation Report Sees Vast Job Loss', reported on fears of 'vast unemployment and social unrest'. Erkki Huhtamo suggests that by the mid-1960s these issues were 'widely debated as markers of a technological transformation which was felt to be shaking the foundations of the industrialized world' (Huhtamo, 1999). Huhtamo might be overstating the case somewhat; there is some evidence that those informed about these issues felt they were exceptions and that ignorance amongst the general public was the general condition. Writing in 1964, Hilton comments, 'this cybercultural revolution ... is affecting the lives of millions of human beings *who have never even heard the new words to describe powerful new concepts*' (Hilton, 1964: 139, my italics). However, Winthrop notes that in more specialist arenas such as within social science and academic disciplines within the academy, around the labour movement, civil rights organizations, on the Left, within science and engineering, and in some policy/government arenas there was informed (and ill-informed) debate, perhaps even a 'zeitgeist in which cybernation is seen as leading society toward a most important and critical juncture' (Winthrop, 1996b: 117).

The Triple Revolution

One partisan report on cybernation that produced heat, if not, according to its critics, light, was the Ad Hoc Committee's *Report on the Triple Revolution*, already introduced. This was written by 'thirty-two prominent social critics and economists' (Block, 1983: 5),[6] a mix of professional, academic, industrial, and labour movement contributors including Alice Hilton (but not Donald Michael). Contemporary readers noted resonances with J.K. Galbraith's *Affluent Society* (Galbraith, 1999, Block, 1983), and Hilton opened her paper, 'An Ethos for the Age of Cyberculture' (1964), with a reference to Galbraith's work. The *Report* was influential in setting out a liberal and left agenda for a post-work world and became a sounding board in cybernation debates. It found support but also provoked some hostility from the Right (and libertarian right), from sections of the organized Left, and from elsewhere. Notably, a theological objection was raised by those feeling the loss of one of the essential components of the fallen human condition – '[the] in the sweat of thy brow shalt thou have earned thy bread labour and sustenance relation now being "cancelled".' (Schwartz, 1966).

The *Report* considered the possible combined impact on American culture and society of atomic technologies and the advent of unwinnable war, computers and 'the cybernation revolution', and the human rights revolution. Their 'simultaneous occurrence and interaction' meant 'radical alterations in attitude and policy' were necessary, but it was cybernation and the end to 'job holding as the general mechanism through which economic resources are distributed' that could reorganize existing programmes for justice (civil rights) and for peace (an end to the nuclear threat), which last was regarded as the 'supreme issue'. The *Triple Revolution* document was globally ambitious in other ways too, linking questions of social justice and rights, and industrial automation and its social consequences arising in the US, to emerging social movements well beyond its borders, for instance.[7] It viewed cybernation as a mechanism that could transform the logics of the market and the priorities of national economies, which latter needed no longer to be 'aimed far more at the welfare of the productive processes than the welfare of people'.

The report acknowledged that cybernation was risky. Its signatories feared violent social breakdown if change was not managed, and were concerned that cybernation without planning could allow 'an efficient and dehumanized community to emerge by default'. But they were optimistic, arguing that 'cybernation, properly understood and used, is the road out of want and toward a decent life'. Hilton thus prophesied 'mutual destruction or a society of affluence and leisure' (Hilton, 1964: 141). Alongside affluence there was also talk of freedom. As the *Report* put it, 'cybernation at last forces us to answer the historic questions: "What is man's role when he is not dependent on his own activities for the material basis of life?"' Echoes of that other *Manifesto* are clearly heard, and it becomes clear that the real prize, as it was in Marx's original, is self-determination. The *Report* declares that '[a] social order in which men make the decisions that shape their lives becomes more possible now than ever before; the unshackling of men from the bonds of unfulfilling labor frees them to become citizens, to make themselves and to make their own history'.

The *Triple Revolution* was published in *Liberation*, presented to President Johnson (March 1964), and read in government circles. Its ideas were discussed in mainstream and specialist media, notably in '*avant garde* periodicals of ideas', including specifically the *Correspondent*, *New University Thought*, *The Minority of One* (Winthrop, 1966a: 117). It was often received with some warmth and awarded serious attention in labour circles and circulated amongst civil rights activists; it was on the curriculum at the Mississippi free school camps for instance.[8] Its propositions also percolated elsewhere; notably, it features in Harlan Ellison's 1967 iconic SF collection *Dangerous Visions* by way of William Jose Farmer's 'The Riders of the Purple Wage', a dystopian take on a future leisure society of staggering violence, banality, and inspiration. The report was unambiguously hated by some. Writing in the autumn of 1966, on Wiener's earlier prediction of imminent disaster, the libertarian Yale Brozen mocked 'amateur social scientists such as Norbert Wiener' for his 1949 predictions of 'a decade or more of ruin and despair' as automation replaced jobs, and sneered that although the catastrophe appeared to have been retimed this hadn't stopped the 'doom criers' of the Ad Hoc Committee from suggesting the time of employment would soon be over because 'men cannot compete with these machines' (Brozen, 1966: 19).[9]

Cybercultural partisans? Key strands in the debate

The Conference on the Cybercultural Revolution was in large part a Triple Revolution show. Many *Report* contributors (including

Hilton, James Boggs, the civil rights and labour activist, and auto worker, Ben B. Seligman), presented or contributed. Some who were involved in the Institute for Cybercultural Research, which formally hosted Hilton's event, were also signatories to the Triple Revolution initiative. These overlaps indicated what Winthrop defined as the 'common convictions of the partisans of the coming age of cyberculture' (Winthrop, 1966a: 115).

The conference took up three key propositions of the Triple Revolution: that cybernation would come about, that it would change the landscape of work and leisure, and that this would have deep social consequences for the present and future. Fierce debates arose around the extent and timing of disruption, how it might be managed, who the beneficiaries and losers might be, what a new society might look like, and how it might be ordered. A broader range of voices from the organized Left, labour organizations, academia, and industry were heard, and also present were people from the new computer industry, some Washington policy specialists, and at least one feminist – Betty Friedan was a discussant (Hilton, 1966).[10] Finally, there was the presence of Arendt, on the one hand, and of Donald Michael, on the other. Arendt's arguments are returned to in the final third of this chapter, but first the broad stakes of cybernation as they emerged at the conference are laid out.

Work

The central issue was the revolutionary proposition that the means to live would soon be divided from the need to work. The implication of this shift, mass or majority unemployment in very short order, which was the *Silent Conquest* prediction, galvanized debate. The prospect of breaking this most basic connection, variously taken as a biblical injunction, as the previously natural order of things now to be overturned, or as a particular industrial and political arrangement, was returned to repeatedly. It was framed as a demand (by some Ad Hoc Committee members), and variously as an intractable or resolvable problem; schemes for Living Certificates, essentially forms of Universal Basic Income, were invoked, for instance.

Dissent came from some ideologically wedded to work, seen as the locus of the humanity and dignity of the individual 'man' (almost always the 'man'), and as central, in the form of organized labour, to societies' hopes for more just futures. Notably, Victor Perlo, CP-aligned leftist (and hero of the McCarthy hearings), was adamant that cybernation could produce new forms of work because wants and needs would increase with an acceleration in the capacity to produce. He predicted new areas of exploration, given peace, 'under the earth and water, in space, in the biological health area ...' (Perlo, 1966: 225),[11] and looked forward to jobs not of the 'grim old-fashioned kind',[12] which might include computer programmers, astronauts, and development workers[13] (Perlo, quoted in Hilton, 1966b: 236). Some dismissed this as wishful thinking. Maxwell Goldberg declared: '[T]he history of the interplay between technology and human labor is the history of the progressive displacement of such labour by labor saving devices invented by humans ... the rest is commentary (Goldberg, 1966: 154).

Timing

The timescale for widespread cybernation was disputed. Various presenters invoked economic analysis, presented empirical evidence (often disputed), or offered speculation based on expert knowledge in relevant domains – notably the nascent technical industries. Some identified complacency around the likely rapidity of change, others argued that the velocity of cybernation was exaggerated; Paul Armer, of RAND, identifying himself as rather cautious, said the 'cybernation of all production' would happen, but added that men would still be working for 'at least two decades' (Armer, 1966: 249).

Transition

The conference explored cybernation and immediate disruption, and the transition to a fully cybernated society both together and apart. Some who felt full unemployment was unlikely to arise in the near future argued for a response framed within existing categories – around labour rights, or rights to work, and so forth. This included demands for large-scale investment in public work programmes. New schools, hospitals, and roads would create jobs, soaking up the newly unemployed and enabling society in general to reap the benefits of automation. Tensions are evident between those considering cybernation's effects on entire populations, and those considering it in relation to specific social groups and classes – the traditional working class and organized labour, Black people, and women. The various perspectives of course depended on the situated position of the speakers, and on their political commitments; Perlo's position, for instance, was founded on a belief that 'capitalism and not the machine' causes mass unemployment. His calls for a planned economy and a 'guiding of public interest by the builders of that life rather than (by) the aristocracy of wealth' (Perlo, 1966: 222) are consistent with that. Accepting that cybernation would produce massive disruption, he nonetheless argued that work neither 'will' nor 'should' become obsolete 'this century'[14] and was one amongst a group at the conference demanding acknowledgement of the urgency of the day over speculation about possible futures (Perlo, 1966: 237). The African American activist and auto-worker James Boggs,[15] an Ad Hoc signatory of the *Triple Revolution* report (Boggs, quoted in Hilton, 1966b: 20), was for full cybernation and an end to wage labour. He argued that rising unemployment would be felt first by those whose experience of labour was of recent and brutal exclusion, those who were already discriminated against in the workforce; African Americans being 'last in' to many of the jobs in organized labour, and then taking on 'scavenger' roles, would be first out (Boggs, 1966: 167). For Boggs, the dignity of labour, even labour itself, is less compelling when such labour does not offer benefits – or generate solidarity, without it having to be fought for. He thus argued that it was 'wrong that so many people think that man must labor and be punished in order to be permitted to exist', adding, pointedly, that it was no longer necessary for man to be a slave (Boggs, quoted in Hilton, 1966b: 243–244).

Boggs nonetheless feared the violence of transition. He wasn't alone. The philosopher and activist Grace Boggs also feared unrest, whilst James Houghton of the National Association for the Advancement of Colored People invoked the 'real and immanent possibility of black and white workers fighting each other for jobs that are rapidly disappearing' (quoted in Hilton, 1966b: 237). The difficulties of transition, and the virtual inevitability of the uneven distribution of the pain it would bring about, were key concerns of Donald

Michael, expressed in the *Silent Conquest*, and in his paper and discussant interventions at the conference. These concerns about transition led some – including James Boggs – to advocate very rapid change. Hilton too argued that it was important 'to complete cybernation as quickly as is technically possible' (Hilton, 1966b: 152), to reach 'the age of cyberculture' (Hilton, 1966b: 150). This, she said, would not solve, but might 'resolve', extant social issues. Her argument has striking parallels with contemporary accelerationist appeals to get out of trouble fast; it wasn't only Hilton's fear of the immediate threat of a war from which 'the earth may never recover …' (Hilton, 1966b: 149), but a desire to accelerate away from existing contradictions and social tensions that founded her position. The focus on possible futures was, however, regarded as escapist by some, Perlo deriding those 'taking refuge in talking of solutions for the distant future' and calling for concentration 'on the tough questions now' alongside discussion of the 'longer range' but still proximate programmes he had identified (Perlo, quoted in Hilton, 1966b).

The future society

Despite objections, a large part of the Cybercultural conference *was* devoted to exploring the possible future shape of an 'Evolving Society', and the burning question was how a 'non-economic life' might be lived. Leisure, politics, the arts, ways of living, and ways of being human, the latter including issues of memory, intelligence, happiness, and the good life – were all chewed over. Three striking features of this debate, to which Arendt responded, are now explored.

Cultural renaissance figured largely. Harry Perks, a Californian systems specialist, felt the 90 percent freed up from work would produce 'the goods of civilization … art, science, literature'. His peroration was a declaration for change: 'I do not fear the leisure society. I am impatient for its arrival … livingry as an alternative to "overkill"' (Perks, 1966: 196, 198). Hilton claimed the leisure society would see the flowering of new forms of cultural activity. To which she added, with a nod to Arendt, a hope for the flowering *also* of new forms of politics. To bring these things about many felt that a key would be education, not only to educate the newly freed society 'into culture' but also to bring 'the people' into new forms of considered citizenship. Some wondered how the newly unemployed would use their time, and if they would find it productive in new ways. James Boggs argued that African Americans were well positioned for a cybernated society, already having experience of being thrust out of the working day. Betty Friedan invoked the experience of middle-class women and their lives of more or less involuntary 'leisure' in the (newly technologized) home, and called for women to be allowed to fully share the benefits of universal cybernation. Her characterization of the fate of women as the *involuntarily* cybernated (it turns out the problem *does* have a name) possibly reveals her ambivalence to a society beyond work (Friedan, quoted in Hilton, 1966b: 317–318). Friedan, arguing for women's perspectives to be taken into account, also noted that 'most of you ... are men ... most of those who think about the cybercultural revolution are men' (Friedan, quoted in Hilton, 1966b: 62).

A rather different debate centred on the post-human. In a euphoric contribution Ted Silvey of the AFL/CIO (American Federation of Labor and Congress of Industrial Organizations) envisaged the computer as 'a giant cerebellum' on the 'outside of the human head' taking care of billions of details to release the 'higher elements of the [human] brain'. This vision remained rather human in the end, since it was said that cybernation would in this way begin the realization of 'man's highest glory' (Silvey, 1966). Others were more fearful. If the prize was a new and more positive existence, the fear was that 'sentient man' would be replaced by the 'compleat robot' (Seligman, 1966: 166).

The third feature of this debate was the marked *lack* of critical considerations of consumption; for instance, around the constitution of artificial as well as real need in the future. The leisure society was often envisaged as a society of plenty, in which all would be rich, and abundant goods would be freely available. For some this vision was almost pastoral: one bucolic vision of the cybernated life was of rush baskets of goods picked up by happy humans from the end of robot production lines. Old forms of life would continue but some roles within them would be undertaken by machines; the ease with which the roles to be replaced are named in raced terms (mechanical 'slaves' would replace human labour) is striking.

Many were hopeful. If the end of work could be achieved without bloodshed and breakdown, an engaged society in which culture flourished, in which populations thrived, in which life could be good for all, and the future glimpsed on the far horizon even better, could emerge. This leisure society utopianism was tempered for some: Edith Goodman, a discussant, wondered if post-cybernation life might be as empty as the crowds happily queuing for heaven at the World Fair (Hilton, 1966b: 240).

Arendt and leisure

Arendt's conference paper 'On The Human Condition', accepts cybernation as a fact, sets aside all matters of transition, and is marked by a vision of the world beyond work that is very bleak indeed. 'On The Human Condition' does not dwell on the probable injustice of transition into a cybernated society for already discriminated-against groups, raised in other papers. Arendt indeed adamantly refuses to deal in special pleading or to take up sectional or intersectional issues. She will not be sidetracked. There is no response offered to James Boggs' concerns over the plight of those African Americans 'last in' to organized labour and likely to be first out. Arendt instead sets out to 'pose the problems from the point of view of the average citizen of the United States of America, *not members of any specific class of the population*' (Arendt, 1966: 214, my italics).

The paper constitutes an utter rejection of romantic idealizations of a post-work society in which new forms of cultural life, and new forms of exalted humanity, may arise. Visions of art and poetry are scythed through. Arendt refused to offer her audience any hope for a utopian society beyond the transitional period so many worried about. Instead she argued that the presumed felicity of a life of cultured leisure freely consumed in a society beyond work was a chimera. Life for humans under conditions of full cybernation could well be an empty, and endless, void. There is nothing automatically uplifting about the end of work; 'we must not think that ... culture can just happen – when there is "free" time' (Arendt, 1966: 217).

The argument begins with a consideration of 'intellectual activity, as such', with rising machine capacity, and with the possibility that machines could take over activities 'identified as activities of the human mind' (Arendt, 1966: 214). Noting developments in computing, Arendt invoked chess computers as an example of the growing brain power of computers. Conceding that chess acuity might be a measure of human intellectual acuity, she argues that it cannot be a measure of somebody's qualifications 'as a human being'. If machines can do something that has hitherto appeared to mark specifically human intelligence, then, says Arendt, the marker must be changed for the sake of the 'dignity' of humans (which is not at all the same as the dignity of labour); or, as I read this, in order that the human condition might continue to be recognized in its uniqueness in conditions of machine encroachment on formerly uniquely human capacities.

Arendt also briefly discusses the limits of computer memory, commenting that humans also have a capacity to forget and to adjust, even to the loss of another, but that these two forms of forgetfulness are not the same. Whilst computers have storage capacity, humans have a capacity for remembrance which does not come from a pre-existing substrate but is *made* (my italics) through action in the world, or in the public realm (see also Arendt, 1998: 208). This has nothing to do with media storage, or 'technical memory' (Arendt, 1966: 215), indeed there is nothing of the machine about it. What come to be memorialized by humans are memorable actions, or worldly interventions. Arendt's argument is that the human capacity to engage or act in the world – and the distinctively human capacity for remembrance is only one part of this – is threatened by full cybernation and the world it may bring about.

She reads cybernation as a new development in transformations in the active life of humans over time. Here it is useful to turn to distinctions drawn between labour, work, and action, as components of the active life (*viva activa*) in the *Human Condition* and to the definition of action in its properly political moment. A fully active life is distinguished from labour (the sweat of the brow) and making (work undertaken on an object). The Industrial Revolution has already replaced the craftsman or *homo faber* (who works) with the socialized labourer distanced from tightly coupled circuits of labour and reproduction, and without the 'animal' compensations this could offer. Humans are already, and more than ever before, distanced from the world – in refusing the valorization of labour Arendt parts company with Marx.

In the conference paper Arendt argues that cybernation might produce a further shift, and one more radical still. Entirely releasing humans from their alienated labour would intensify their alienation – the term here implying an estrangement from any form of earthed or engaged life, an estrangement from that which *conditions*. This is explored first through adaptability. Arendt invokes her own experience of technologically driven change – from 'horse drawn carriages' to 'credit cards' in one lifetime, to consider adaptability and what it needs. Humans, she declares, are adaptable because of how they are 'conditioned by the world around them, the world in which, and with which they *engage*'.[16] This 'feed-back' cycle is quite obvious, she says: 'once the environment has really changed we are conditioned, even though we may know very little about the conditions that have moulded us' (Arendt, 1966: 218).

Arendt's fear is that human adaptability will fail in a fully cybernated society.

Leisure/idleness

The reason is the nature of leisure society at the end of work, which Arendt, drawing on Hilton's term,[17] explores in terms of 'idleness', something which '*is* ugly … [it] frightens us a little' (Arendt, 1966: 217). Lenz (2005) and Postel (2002) note Arendt's sense that leisure is for stopping and thinking. A clear space is needed for the task of engaging in a 'cooperative attempt to shape the world' (see Hilton, 1966b: 136). Against the Romans, who dealt with the idle life through bread and circuses, Arendt sets the Greek ideal of the citizen 'whose political tasks were so time-consuming he had neither vacant time, or idle time' (Arendt, 1966: 217). The busy cultural activities envisaged as central to a post-work society will no more deliver this space for leisure than what Hilton, exposing her class privilege, disparaged as activities designed to fill idle days – beer and television – for those living almost completely private lives (Hilton, 1966b: 149). For Arendt, hopes that a new society would remain 'active', that 'many creative activities and interests commonly thought of as non-economic would absorb the time and the commitment of many of those no longer needed to produce goods and services' (the aspirations of the Triple Revolution), were thus *beside* the point, or were working with an entirely different sense of what might constitute the active life.

Why could leisure *not* be used in precisely the pursuit of the political activity of which she talks? In her paper Arendt considers the 'laborless strata of human society' (Arendt, 1966: 217). She remarks on the difficulty traditionally privileged groups have found in handling leisure, 'afraid of deteriorating in their complete freedom' (Arendt, 1966: 218). Given this difficulty, the aspiration (aired at the conference) that freed from labour all shall 'be able live on so high a level' seems to her unrealistic. She appears to believe that there is nothing to be learned from the experience of groups systematically excluded from work. To Friedan's intervention, 'Half the population is now living in the cybercultural era ... Will you let women in?', and to James Boggs and others fearing an immediate future of increased race discrimination and violence, but also arguing that African Americans have experience of precarity, she has little to say.[18] Certainly, she does not align with those who see the leisure society as a form of restitution for past injustice or pain.

Arendt's argument is that leisure is required for properly political action but is neither easeful nor easy to handle. A world beyond labour and work, in which the vast majority are condemned to idleness, will condemn that majority to an existence filled with 'vacant time' (Arendt, 1966: 217). Worse, in a cybernated world this may be a permanent condition, since in this vacancy, or vacuum, this place neither offering labour nor leisure, there are no conditioners, no traction, no gears to engage or call forth action, or enable adaptation, not even the residual forms of engagement found in earlier forms of industrial labour. Public life is stripped away. The world, which conditions, which produces adaptability, and which provides grounds across which action is called forth, disappears; '*For vacant time is not a conditioner. Vacant time is nothingness*' (Arendt, 1966: 218, my italics). In *The Human Condition* work is done to 'deconstruct the aura of necessity that surrounds all thinking about labour', argues Lenz (2005: 139); the conference paper makes clear that, despite that, labour is necessary – in that it is what remains of the *viva activa*. Without it there is nothing (again a vacuum). This is why, for Arendt, this life without work may be a move into a sort of death, an *inhuman* condition. In these circumstances it is hard to envisage how something new may emerge or be born. It is possible that the human species, adaptable though it has always been, may not be able to adapt to vacant time.

There is a strong commitment in Arendt's work to the idea that 'action is open to all people', being 'rooted in the condition of being born' (Arendt, 1998: 86), and being therefore a universal possibility (Young-Bruehl, 2006). Politics is defined as action, as 'a practical project, a shared concern of citizens, and Firer Hinze notes Arendt's 'passionate commitment to the concrete' (Firer Hinze, 2009: 29). Many at the Cybercultural conference had a different sense of practical politics. The response to Arendt's line of thinking suggests that they were at once beguiled and repulsed by her views. The perceived elitism of Arendt's discussion of who could 'handle' leisure certainly disturbed. Even Alice Hilton noted, rather dryly, that leisure was apparently always one of those things that somebody else couldn't handle. Arendt's refusal to take on board the sectional claims and appeals of particular groups also nonplussed some, although it is consonant with Arendt's divisions between the properly political realm and the social sphere.

Parallels might be drawn with Arendt's controversial intervention around desegregation battles in Little Rock,[19] where she criticized parents of the school children attending White schools for mistaking the proper grounds of action. Making this intervention, she presumed her own underlying sympathies and affiliations were clear, when for many her comments placed her on the wrong side of the fight (see Firer Hinze, 2009). Anne Norton comments that 'in relation to Little Rock, the uneasy fit between her writings on race and her disavowal of complicity in an unjust racial order shows the danger of taking one's sympathies for granted'. (Norton, cited in Gines, 2009: 73). Perhaps at the Cybercultural conference Arendt also took her general political orientation as read and, as in 'Little Rock', this wasn't necessarily as clear as she believed. But the parallel to be made is also that in both cases what is at issue (for Arendt) is the relationship between the sphere of politics and the sphere of society, the latter defined as 'that curious, some-what hybrid realm, between the political and the private in which, since the beginning of the modern age, most men have spent the greater part of their lives' (e.g. Arendt, 2003: 205, cited in Firer Hinze, 2009: 46).

Arendt's fear of the leisure society is that this hybrid realm is all that there is. For her there is nothing in the life of leisure arriving in the wake of cybernation, a transformation dealt with not in economic terms but essentially as a coming ontological condition that is going to make it easier to act, or to act politically, or become or remain, fully active citizens. Instead, ease is part of the problem. Arendt writes in *The Human Condition* that:

> The easier that life has become in a consumers' or laborers' society, the more difficult it will be to remain aware of the urges of necessity by which it is driven, even when the pain and effort, the outward manifestations of necessity, are hardly noticeable at all. The danger is that such a society, dazzled by the abundance of its growing fertility and caught in the smooth functioning of a never-ending process, would no longer be able to recognize its own futility – the futility of a life which 'does not fix or realize itself in any permanent subject which endures after [its] labor is past.' (Arendt, 1998: 135)

This is where memory (human, collective memory) and the *viva activa* coincide. We might see this lack of capacity to make history as a form of presentism.

Exergue: the human condition and technological condition of the world

Arendt is widely read as a political theorist and a critic of modernity (see e.g. Canovan, 1998: vii–viii). Consonant with this the classification of the activity of life through the categories of labour, work, and action, and the examination of changing hierarchies between them as they emerge in the mid-20th century constitutes the central material of *The Human Condition*. But what is the place of the technological in *The Human Condition*? In the work's critical reception certain registers through which technology is taken up have been rather neglected. In Young-Bruehl's account *The Human Condition* is subsumed into Arendt's larger body of work on political totalitarianism. Others also note that technology and totalitarianism are tightly ('organically') linked in many of Arendt's works (Canovan, 1998: xi). Attention is thus drawn to Arendt's thinking on technocratic rationality as that which might tend towards destroying politics, and this might be a definition of totalitarianism *as* a form of government (and see Young-Bruehl, 2006: 39). Less attended to is Arendt's more *specific* consideration of questions concerning technology and technological innovation. Symptomatically, there is no entry for technology in the index to Benhabib's authoritative edited collection on Arendt, *Politics in Dark Times* (Benhabib, 2010: 4).

In her introduction to 1998 edition of *The Human Condition* Margaret Canovan noted the dialectical relation it entertains between two 'strangely unconnected' events: the flight into Space and the advent of automation (Canovan, 1998: x). It begins with Sputnik's rise and ends with the invention of the telescope, a rather beautiful inversion of archaeologically feasible causality in itself. Considering these technologies and connections that are drawn between them prompts a reading of *The Human Condition* as an *account* of technological change, which is not quite the same as reading Arendt as a philosopher of technology (see Melis Bas, 2013), but is rather at odds with accounts that subsume questions concerning technology into questions concerning technocratic rationality only.

To start, then, not with rationality but with action: Arendt argued that the *viva activa* rises with the modern age in relation to the event of technology, that is, the invention of the telescope, an event that does not stand squarely inside the modern age but is at its opening. This event came quietly, as compared with the 'spectacular' discovery of 'unheard of continents and un-dreamed of oceans' (Arendt, 1998: 249), and was less initially disturbing than other events (the Reformation). However, the 'tentative steps' it enabled towards the discovery of the universe increased in momentousness and 'speed', eclipsing the enlargement of the Earth and its subsequent shrinkage through mapping processes (Sputnik and the view from Space comes back in). The telescope offered 'demonstrable fact' where before there had been inspired speculations, those of Bruno, notably (Arendt, 1998: 260). Paradoxically, this produced not exaltation but doubt – Cartesian doubt, addressed by a new reliance only on what the self can know. But 'it was not reason but a man-made instrument, the telescope, which actually changed the physical worldview, it was not contemplation, observation and speculation which led to the new knowledge, but the active stepping out of *homo faber*, of making and fabricating' (Arendt, 1998: 257–263). Thus, if there is a growing reliance on self-made entities of the mind, mathematics, and knowledge that may be tested through more doing (the experiment), the real reversal, says Arendt, is between thinking and doing, and where thinking (philosophy) continues to demand self-inspection, the latter always demands *more*. The alternatives offered are 'redoubled activity' or 'despair' (Arendt, 1998: 293).

This furnishes a consideration of the rise, within the *viva activa*, of the *Homo faber*, or making man, and then the defeat of the *Homo faber*, or the general ascendency of labouring – giving *homo laborans*. *Homo faber* involves an engagement with the world through objects, but there may be a moment of contemplation within the process of making. As Arendt puts it, the arms fall and the 'beholding of the model' takes place (1998: 303). This, the craftsman's differently 'contemplative glance', she says, came to be known by many even whilst the other form of contemplation, that of philosophy, was known only by the few. But even this form of contemplation is said to be at risk.

The trajectory here follows two lines. First there is an exchange in priorities so that the concept of process rises over the concept of being (Arendt, 1998: 296). Eventually fabrication is understood entirely in terms of the former, and in this move, from 'product' to the 'process' (Arendt, 1998: 304), contemplation loses its position in the *viva activa* and in 'ordinary human experience' (Arendt, 1998: 304). Second, in this new world the principle of labouring expands out of the realm of the reproduction of life to become far more widely operational. Labour brings with it the same unworldliness, or privatized existence that it had when confined to the more strictly or literally private life of reproduction. For Arendt this is unworldly because it presumes that what humans share is not the world, but shared or same nature. Moreover, when labouring, intrinsically unworldly, rises in this world, it brings with it a Benthamite ratiocination of measured happiness that is some distance from the forms of making that have their eye on the object made (and in this way may also contain a moment of contemplation), and is even further from action – which would not be measured or gauged in such a way at all.

126   Anti-computing: dissent and the machine

This is accelerated (as part of processes of automation, effectively) when the maker becomes the maker of tools-to-make-tools 'who only incidentally produces things'. By now, judgement is made neither in the realm of the political nor even in relation to things and their usefulness, but concerns only process, or the 'amount of pain or pleasure experienced in the production or consumption of things' (Arendt, 1998: 309). As Arendt puts it, the modern age thus operates under the assumption that 'life and not the world, is the highest good' (Arendt, 1998: 318); or, as Lenz argues, Arendt's critique of the labour society is 'based on a criticism of the absolute domination of labour at the expense of other form of activities' (Lenz, 2005: 145).

Beyond the last stage? The deadliest passivity

The ways in which this transformation occurs, and the glancing invocation of Bentham, might suggest how the history here, of the transformation of work into labour and of the excision of action so that – to repeat – life not world is the highest good, connects to the framing trope of the telescope, and in particular to its operation as something that may reposition humans. Arendt heads the final chapter of *The Human Condition* with an epigraph from Kafka: 'he found the Archimedean point, but he used it against himself; it seems he was permitted to find it only under this condition' (cited in Arendt, 1998: 248). The point is clear. This new telescopic perspective, at once universal rather than worldly, and individualized, or operating with principles of personalization, rather than dealing with men and women, is at once newly intimate, swallowed and held within, *and* radically external. It is tempting to read this as a peculiarly far-sighted Arendtian comment on contemporary big data and personalization processes today, but the germane point here is that this condition is *technologically* given or made.

Arendt also provides a brief commentary, a matter of a couple of pages (Arendt, 1998: 322–323), on what might lie *beyond* the shift into generalized labour. Essentially, widespread automation is here under discussion, although there is none of the matter-of-fact certainty that full automation/cybernation *will* come that is so startling in the conference paper. Assessing the contemporary moment, Arendt says one element of the active life, labour, has become 'victorious' but is itself in its final stages. Labour has become too 'lofty' a term to describe the circuitry of consumption/job holding characterizing the contemporary world: 'we have proved ingenious enough to find ways to ease the toil and trouble of living to the point where an elimination of laboring from the range of human activities can no longer be regarded as utopian'. Her fear is that soon all that will be left is consumption. But she adds that already 'the society of job holders' demands of its members 'a sheer automatic functioning, as though individual life had actually been submerged in the overall life process of the species and the only active decision still required of the individual were to let go ... acquiesce in a dazed, "tranquilized" functional type of behavior' (Arendt, 1998: 322).[20]

Functional behaviour here is a response that is, in Arendt's terms, highly unpolitical, not only because unreflexive but also because no longer individual but only individuated. She thus argues that 'the trouble with modern theories of behaviorism is not that they are wrong but that they could become true ...' (1998: 322). This clarifies the discussion of conditioning and adaptation in the 'On The Human Condition' paper, where individual adaptation to changing (technological/techno-social) environments in which humans find themselves is replaced by the automatic modulation of humans, from the outside. There, we will recall, Arendt argued that it was 'quite conceivable that the modern age ... may end in the deadliest, most sterile passivity history has ever known' (1998: 322).

The perceived threats here overlap. On the one hand, the fear is of the rise of a society of labourers with nothing to labour on. On the other, there is the automated functioning of the human on automatic pilot. Finally, there are those technologies of vision and perspective through which this, the human body, the body now unpolitic, might be viewed, and manipulated. Arendt's final – and startling – warning in *The Human Condition* is thus that some possibility for action remains, but is often the prerogative of the scientist.

The rest of us, or the world in general, have a new perspective but can only watch ourselves, as if we were processes, so that the modern motorization, for instance, 'would appear like a process of biological mutation in which human bodies gradually begin to be covered by shells of steel'. Moreover, from the perspective of 'the watcher from the universe' – and this would now be the perspective of unearthed man – 'this mutation would be no more or less mysterious than the mutation that now goes on before our eyes in those small organisms ... which have mysteriously developed new strains to resist us' (1998: 323).

Alice Hilton told the Cybercultural conference that 'The cybercultural society is a cybernetic system. A cybercultural society is as great as the universe' (1966: 340). What we understand at the end of *The Human Condition* and at the end also of the Cybercultural discussions[21] is that Arendt is hostile to the universal viewpoint, at least in so far as it obliterates men, humans, replacing them with Man, who is himself at once singular, and also singularly ill-defined, no longer worldly but part of the pattern of the Earth as seen from the inhuman view of Space.

There is suddenly here something from Heidegger in Arendt, as she too points to how man is caught and remade in the technologies that seem to give him mastery. The Archimedean point, the point from which it is possible to move the Earth, has been achieved, but its positioning, at once in Space, and, as in Kafka's example, in the body, tends to produce conditions in which humans are no longer able, despite their own role in the making of this technology, to act.

Arendt was conscious of forms of technological evolution and development. What the 'On The Human Condition' paper makes clear, elaborating and focusing on some of the arguments begun in these final pages of *The Human Condition*, is that she is *specifically* informed, and engaging in debate on the matter of cybernation, the end of work, due to the influences of computerization and cybernetics. We might say that she had a take on this. And it was hostile: Arendt was anti-computing.

Arendt's explicit engagement with cybernation was brief. By 1970, in *On Violence*, she would comment that if one motivation for student protest was 'the simple fact that technological progress is leading in so many instances straight into disaster' (cited in Young-Bruehl, 2006: 149), she also added that the real issue was not unemployment. Rather, 'the seemingly irresistible proliferation of techniques and machines, far from threatening certain classes with unemployment, menaces the existence of whole nations and conceivably of all mankind' (Arendt, 1970). The issue remained politics and peace, but now the focus was not on technology's role in labour and its ending.

After the event

There was no cybernation revolution. What was briefly taken seriously by government, industry, sections of the Left, and civil rights movement, nascent feminist organizations, and Arendt herself, became afterwards a 'scare', as Ganz, who was there at the conference with the other partisans of cyberculture, put it (Ganz, 1966).[22] This wasn't only about practical barriers to real implementation but also a matter of desire and support; the cybernation imaginary faltered. What happened? In 'The Selling of the Productivity Crisis' (1983) James E. Block asked why public discourse 'led away from the consideration of a society less centred around the workplace'. He blamed 'entrenched interests, who wish market inequalities to persist, and do so by shifting the blame onto workers', and the 'deep collective failure' (of the Left) to confront uncertainties (partly because of its historical association with the working poor), and said that as a result 'discussions on automation, non-work society, and alternative forms of distribution held in the late fifties have been deferred for a generation' (Block, 1983: 13).

Cyberculture to precarity?

After cybernation, analysis of the consequences of rising levels of computerization within society, and of shifts in computing itself (from brute automation to refined control, from ungainly giants to office machines, from rarity to proliferation), took other turns. Daniel Bell's later work on the post-industrial society as a knowledge society, is key here. Essentially, as an intervention at odds with the cybernation thesis (see e.g. Bell, 1965, Michael, 1965) it won out (Bassett and Roberts, 2020).

A different perspective on the post-cybernation debate is given by tracing out the notion of the cybercultural itself. If her conference paper usefully interrogates Arendt's own thinking on leisure and work, at a very particular moment in time, her contribution to the debate on 'cyberculture' contributed to how the term was understood critically, back then, on this, one of its very first outings. Cyberculture at the conference was bound up with questions of radical change in the political economy of the US, with the questions of war and peace, as well as with the civil rights movement. Cybernation did not occur, and the term itself was dropped. By contrast, whilst the term cyberculture was also eclipsed in the aftermath of the scare, it *did* come back. But when it returned in the mid- to late 1980s in response to developments including the internet and its associated imaginaries, it did so in a peculiarly etiolated guise – being generally taken to refer not to questions of labour, work and action, but to the formation of contemporary social subjectivities: cyber-lives for cyber-selves, and cyber-society.

Mainstream 1990s cyberculture was perhaps still interested in questions of labour and leisure, but now in the fusion of the two; it celebrated or excoriated a new playground. Its apogee, largely imaginary in that the technologies that were supporting it were nascent, was virtual reality, seen both as the other of cyberspace and as its fully realized form. In the latter it was roundly abused by commentators, including notably Kevin Robins, who, in terms Arendt might have approved of, accused it of turning its back on the world we live in (1996). The term largely slipped out of use in the late 1990s and is now remembered as painfully 'of its time', as William Gibson noted.[23]

Huhtamo discerns a trail leading from cybernation to interactivity, to being 'in' the media, perhaps. I would lay a different trail; this would move from cybernation to precarity[24] as it is emerging in relation to new developments in automation. Arendt's thinking lets us see that this condition needs to be explored not only in terms of political economy but also in relation to what might be termed a political ontology. Critiques of her thinking, not least those made by the Cybercultural conference audience she seduced and abandoned, would suggest doing so not only in relation to the human condition but also by questioning the homogeneity, the boundedness, the unacknowledged *situationality* of position, of 'the human' in that term.

Notes

1 19–21 June, 'The hottest days in any June', Hilton (1966b: Foreword).
2 A facsimile of the conference programmes can be seen at www.fredbernardwood.org/LifeTimes/TechSocietyIBM/ConfCyberRev64.pdf.

3 The long, hot summers of 1964/65 reference civil disturbances/riots. Harlem in 1964, Watts in Los Angeles in 1965, Detroit in 1967. The Civil Rights Act (1964) was signed by President Johnson following Kennedy's assassination.
4 Great Society legislation conceded basic civil rights. The context was also cybernetics and the rise of computerization. The year 1964 saw the US launch of IBM System/360, the first minicomputer, built by Digital Equipment Corporation (DEC). See e.g. IBM's history pages, www-03.ibm.com/ibm/history/history/decade_1960.html .
5 Arendt's speech is gathered in her papers at the Library of Congress, filed under organizations 1943–1976 'Conference on Cybernetics, New York'. There is also a brief typed correspondence from Alice Mary Hilton asking for alterations.
6 Signatories to the Triple Revolution included Alice Hilton, James Boggs (who also wrote a special issue of *Monthly Review* on the 'workless society'), and multiple Nobel Prize-winner Linus Pauling. Another context to the Evolving Society debates was various proposals for organizing exchange in a society beyond full employment. These included proposals for Living Certificates.
7 E.g. Rights in the US were deemed 'only a local manifestation of a world wide movement'.
8 'In the summer of 1964 over forty Freedom Schools opened in Mississippi, part of Freedom Summer, a project of the Southern Civil Rights Movement. The goal was to empower African Americans in Mississippi to become active citizens and agents of social change' (Welcome to the Freedom School Curriculum Website! written by Kathy Emory, www.educationanddemocracy.org/index.html, www.educationanddemocracy.org/FSCfiles/C_CC2a_TripleRevolution.htm). The March 1964 curriculum conference included segments on 'The Power Structure' by SNCC research director Jack Minnis, and on the history of the Freedom Movement. There were also *Liberation* magazine reprints on the 'Triple Revolution'. See Staughton Lynd's *The Freedom Schools, An Informal History* (accessed February 2019), www.solidarity-us.org/current/node/477.
9 Brozen was invoked at the conference for his gradualist views on cybernation, by Maxwell Goldberg (1966: 156).
10 In my interview with Alice Mary Hilton, she said she had no recollection of Friedan attending, suggesting that she had been invited to speak but had declined. The Proceedings, however, do record Friedan's intervention, and list her as a discussant citing the 'Feminine Mystique' (Hilton, 1966b: 392).
11 Panel discussion.

12 W.H. Ferry in discussion, Hilton, 1966b: 229.
13 Hilton agreed with the policy of building, but thought it wouldn't provide jobs in a cybernated era. For her the prime purpose of technology was the 'disempowerment' of human beings, this latter potentially for the good (ES: 233).
14 W.H. Ferry (VP Fund for the Republic), a discussant, attacked Perlo's reliance on jobs as crude economism and 19th-century thinking: 'any plan for a civilized society must ... include plans for jobs.' He argued that Perlo should bring his 19th-century thinking up to date. (Hilton, 1966b: 229).
15 Best known for authoring *The American Revolution: Pages from a Negro Worker's Notebook* in 1963, James Boggs worked at Chrysler from 1940 until 1968. He was the husband of Grace Boggs, also an activist.
16 Hilton says 'adults too must be taught', but focuses on evidence of the 'adaptability' of man'. This includes human capacities to learn by forgetting. As she puts it, 'man has always been able to forget the old and learn new ways' (Hilton, 1966b: 135).
17 'Idleness ... the passive endurence of vacant time' (Hilton, introduction to Evolving Society session in Hilton, 1966b: 256). Vacant time was a recurring discussion at the conference.
18 The recompense is the hope that the freedom to act would produce a revitalized public realm in which women act. As Dietz (2002) puts it, thinking about Arendt's project more broadly, 'Arendt is offering a way to not look inward, but to value all voices in the public realm ... In Arendt's existential analysis [...] there is nothing intrinsically or essentially masculine about the public realm, just as there is nothing essentially feminine about labouring in the realm of necessity.'
19 See Arendt in the preface to 'Reflections on Little Rock' (2003): 'As a Jew I take my sympathy for the cause of the Negroes as for all oppressed or under-privileged people for granted and should appreciate it if the reader did likewise.'
20 See Dave Eggers' *The Circle* for a contemporary treatment of this kind of sensibility.
21 Final panel discussion: Future society reasons for hope and causes for fear.
22 Ganz was from government circles (US Department of Labor). See also Widner's sense of 'consensus ... we are in the middle of the cybercultural revolution'. He was responsible for a senate committee on manpower (Widner, quoted in Hilton, 1966b: 329).
23 There is a widespread presumption that the term cyberculture was invented alongside the worldwide web and William Gibson's coining of the term 'cyberspace'.

24 This trajectory can be traced in terms of resistance movements too. One thread in this might be Grace Boggs, present at the conference, and active in the Detroit movements for many decades.

Bibliography

Agger, Donald G. et al. 1964. [Ad Hoc Committee] 'The Triple Revolution', *International Socialist Review*, 24(3): 85–89.
Anshen, Ruth Nanda. 1966. 'Our Concept of the Good Life: Idleness or Leisure?', in Alice Mary Hilton (ed.) *The Evolving Society. First Annual Conference on the Cybercultural Revolution – Cybernetics and Automation*. New York: Institute for Cybercultural Research: 301–308.
Arendt, Hannah. n.d. 'Cybernetics', lecture. Hannah Arendt Papers, Library of Congress. Series: Speeches and Writings File, 1923–1975
Arendt, Hannah. 1966. 'On The Human Condition', in Alice Mary Hilton (ed.) *The Evolving Society. First Annual Conference on the Cybercultural Revolution – Cybernetics and Automation*. New York: Institute for Cybercultural Research, 213–220.
Arendt, Hannah. 1970. *On Violence*. London: Allen Lane/Penguin.
Arendt, Hannah. 1987. 'Labour, Work, Action', in J.W. Bernauer (ed.) *Amor Mundi. Explorations in the Faith and Thought of Hannah Arendt*. Canada: Kluwer Academic Publishers, 29–43.
Arendt, Hannah and Mary McCarthy. 1996. *Between Friends: The Correspondence of Hannah Arendt and Mary McCarthy, 1949–75*. London: Secker and Warburg.
Arendt, Hannah. 1998 [1958]. *The Human Condition*. Chicago: University of Chicago Press.
Arendt, Hannah. 2003. 'Reflections on Little Rock', in Jerome Kohn (ed.) *Responsibility and Judgment*. New York: Schocken Books, 193–194.
Armer, Paul. 1966. '"Intelligent" Machines', in Alice Mary Hilton (ed.) *The Evolving Society. First Annual Conference on the Cybercultural Revolution – Cybernetics and Automation*. New York: Institute for Cybercultural Research, 113–118.
Bagrit, Leon. 1965. *The Age of Automation*, BBC Reith Lectures, London: Weidenfeld and Nicolson.
Barnes, Robert F. Jr. 1973. 'Review of *Logic, Computing machines and Automation*, by Alice Mary Hilton', *The Journal of Symbolic Logic*, 38(2) 341–342.
Bas, Melis. 2013. 'A Reinterpretation of Hannah Arendt as a Philosopher of Technology'. Master's thesis, University of Twente, The Netherlands.
Bassett, Caroline and Ben Roberts. 2020. 'Automation Now and Then: Automation Fevers, Anxieties and Utopias', *New Formations*, 98: 9–28.

Bell, Daniel. 1965. 'The Bogey of Automation', *The New York Review of Books*, 5(2): 23–25.
Bellman, Richard. 1966. 'Science, Technology, and the Cybernation Explosion', Alice Mary Hilton (ed.) *The Evolving Society. First Annual Conference on the Cybercultural Revolution – Cybernetics and Automation*. New York: Institute for Cybercultural Research, 75–80.
Benhabib, Seyla. 1993. 'Feminist Theory and Hannah Arendt's Concept of Public Space', *History of the Human Sciences*, 6(2): 97–114.
Benhabib, Seyla (ed.). 2010. *Politics in Dark Times: Encounters with Hannah Arendt*. Cambridge: Cambridge University Press.
Block, James E. 1983. 'The Selling of a Productivity Crisis: The Market Campaign to Forestall Post-industrial Redistribution and Work Reduction', *New Political Science*, 4(1): 5–19.
Boggs, James. 1963. *The American Revolution: Pages from a Negro Worker's Notebook*, New York: Monthly Review Press, www.historyisaweapon.com/defcon1/amreboggs.html.
Boggs, James. 1966. 'The Negro and Cybernation', in Alice Mary Hilton (ed.) *The Evolving Society. First Annual Conference on the Cybercultural Revolution – Cybernetics and Automation*. New York: Institute for Cybercultural Research, 167–172.
Bousquet, Antoine. 2008. 'Cyberneticizing the American War Machine: Science and Computers in the Cold War', *Cold War History*, 8(1): 77–102.
Brick, Howard. 1992. 'Optimism of the Mind: Imagining Postindustrial Society in the 1960s and 1970s', *American Quarterly*, 44(3): 348–380.
Brozen, Yale. 1966. 'Automation: The Retreating Catastrophe', *Left and Right: A Journal of Libertarian Thought*, 2(3): 19–32.
Caldwell, William A. 1966. 'The Revolution in Communications', in Alice Mary Hilton (ed.) *The Evolving Society. First Annual Conference on the Cybercultural Revolution – Cybernetics and Automation*. New York: Institute for Cybercultural Research, 49–58.
Canovan, Margaret. 1998. 'Introduction', in Hannah Arendt, *The Human Condition* (2nd edn). Chicago: University of Chicago Press, vii–xx.
Cowan, Ruth. 1985. *More Work for Mother: The Ironies of Household Technology from the Open Hearth to the Microwave*. London: Basic Books.
Crenshaw, Kimberlé. 1991. 'Mapping the Margins: Intersectionality, Identity Politics, and Violence against Women of Colour', *Stanford Law Review*, 43(6): 1241–1299.
Czarniawska, Barbara, and Eva Gustavsson. 2008. 'The (D)evolution of the Cyberwoman?', *Organization*, 15(5): 665–683.
Davis, Angela. 1982. 'The Approaching Obsolescence of Housework: A Working-class Perspective', In *Women, Race and Class*. London: Women's Press, www.marxists.org/subject/women/authors/davis-angela/housework.htm.
Dietz, Mary. 2002. *Turning Operations: Feminism, Arendt, Politics*. London: Routledge.

Dave, Eggers. 2013. *The Circle*. London: Penguin.
Eglash, Ron. 1998. 'Cybernetics and American Youth Subculture', *Cultural Studies*, 12(3): 382–409.
Eglash, Ron, and Julian Bleecker. 2001. 'The Race for Cyberspace: Information Technology in the Black Diaspora', *Science as Culture*, 10(3): 343–274.
Farmer, Philip José. 1967. 'Riders of the Purple Wage', in H. Ellison (ed.) *Dangerous Visions*. New York: Doubleday.
Federici, Syliva. 2004. *Caliban and the Witch. Women the Body and Primitive Accumulation*. London: Autonomedia.
Firer Hinze, Christine. 2009. 'Reconsidering Little Rock: Hannah Arendt, Martin Luther King Jr., and Catholic Social Thought on Children and Families in the Struggle for Justice', *Journal of the Society of Christian Ethics*, 29(1): 25–50. [See 'There is no cheap grace' (p. 36).]
Friedan, Betty. 1963. *The Feminine Mystique*. London: Penguin.
Galbraith, J.K. 1999. *The Affluent Society*. London: Penguin.
Gans, Herbert J. 2009. 'Working in Six Research Areas: A Multi-field Sociological Career', *Annual Review of Sociology*, 35: 1–19. [First published online as a Review in Advance, 25 February 2009.]
Ganz, Samuel. 1966. 'Summary', in Alice Mary Hilton (ed.) *The Evolving Society. First Annual Conference on the Cybercultural Revolution – Cybernetics and Automation*. New York: Institute for Cybercultural Research 328–332.
Germain, Gilbert. 2004. 'The Human Condition in the Age of Technology', in David Tabachnick and Toivo Koivukoski (eds) *Globalization, Technology, and Philosophy*. Albany, NY: State University of New York Press, 159–174.
Gerovitch, Slava. 2002. *From Newspeak to Cyberspeak: A History of Soviet Cybernetics*. Cambridge, MA: MIT Press.
Gines, Kathryn T. 2009. 'Hannah Arendt, Liberalism, and Racism: Controversies Concerning Violence, Segregation, and Education', *The Southern Journal of Philosophy*, XLVII: 53–76.
Goldberg, Maxwell H. 1966. 'The Displacement of Human Labour: Historic Trends', in Alice Mary Hilton (ed.) *The Evolving Society. First Annual Conference on the Cybercultural Revolution – Cybernetics and Automation*. New York: Institute for Cybercultural Research, 153–158.
Goldberg, Steven. 1991. 'The Changing Face of Death: Computers, Consciousness, and Nancy Cruzan', *Stanford Law Review*, 43(3): 659–684.
Goodman, Leo. 1994. *Papers of Leo Goodman, 1913–1982*. A Finding Aid to the Collection in the Library of Congress. Prepared by Melinda K. Friend. Washington: Library of Congress [revised August 2011].
Hansen, Mark. 2004. 'Digitizing the Racialized Body or The Politics of Universal', *SubStance*, 33(2), Special Section: Contemporary Novelist Lydie Salvayre, 107–133. Available from: www.jstor.org/stable/3685406.
Hayward, Herbert Layton. 1966. 'Our Concept of Society: Drifting or Steered?', in Alice Mary Hilton (ed.) *The Evolving Society. First Annual*

Conference on the Cybercultural Revolution – Cybernetics and Automation. New York: Institute for Cybercultural Research, 275–282.
Heims, Steve J. 1991. *The Cybernetics Group*. Cambridge, MA: MIT Press.
Hilton, Alice Mary. 1964. 'An Ethos for the Age of Cyberculture', AFIPS '64 (Spring): *Proceedings of the April 21–23, 1964, Spring Joint Computer Conference*, https://dl.acm.org/doi/10.1145/1464122.1464136.
Hilton, Alice Mary. 1966a. 'The Bases of Cyberculture: Scientific, Technical, Philosophical, Socio-economic, Political', in Alice Mary Hilton (ed.) *The Evolving Society. First Annual Conference on the Cybercultural Revolution – Cybernetics and Automation*. New York: Institute for Cybercultural Research, 7–22.
Hilton, Alice Mary (ed.) 1966b. *The Evolving Society. The First Annual Conference on the Cybercultural Revolution – Cybernetics and Automation*. New York: Institute for Cybercultural Research.
Huhtamo, Erkki, 1999. 'From Cybernation to Interaction: A Contribution to an Archaeology of Interactivity,' in Peter Lunenfeld (ed.) *The Digital Dialectic. New Essays on New Media*. Cambridge, MA: MIT Press, 96–110.
Hull, Margaret Betz. 2002. *Hidden Philosophy of Hannah Arendt*. London: Routledge Curzon.
Jeschke, Sabina, Ingrid Isenhardt, Frank Hees, and Klaus Henning. 2013. *Automation, Communication and Cybernetics in Science and Engineering 2011/2012*. London: Springer.
Jurgens, Jeff. 2012. 'Defining American Democracy Differently', *Amor Mundi*, 25 October, www.hannaharendtcenter.org/?p=8021.
Kersplebedeb, Karl. 2005. Review of *Caliban and the Witch: Women, the Body and the Primitive Accumulation*, by Silvia Federici, *Upping the Anti*, 2 (December), https://uppingtheanti.org/journal/article/02-caliban-and-the-witch-women-the-body-and-primitive-accumulation.
Keyserling, Leon H. 1965. 'Automation ... Reply by Daniel Bell in response to *The Bogey of Automation* from the August 26, 1965 issue', *The New York Review of Books*, 25 November, www.nybooks.com/articles/archives/1965/nov/25/automation-2/.
Kline, Robert. 2009. 'Where are the Cyborgs in Cybernetics?', *Social Studies of Science*, 39(3): 331–362.
Ledley, Robert S. 1966. 'Computing Machines in Medicine', in Alice Mary Hilton (ed.) *The Evolving Society. First Annual Conference on the Cybercultural Revolution – Cybernetics and Automation*. New York: Institute for Cybercultural Research, 93–112.
Leikind, Morris C. 1966. 'The History of Technology: Man's Search for Labour-saving Devices', in Alice Mary Hilton (ed.) *The Evolving Society. First Annual Conference on the Cybercultural Revolution – Cybernetics and Automation*. New York: Institute for Cybercultural Research, 23–34.
Lenz, Claudia. 2005. 'The end of the Apotheosis of "Labour"? Hannah Arendt's Contribution to the Question of the Good Life in Times of

Global Superfluity of Human Labour Power', trans. Gertrude Postl, *Hypatia*, 20(2): 135–154.
Lynd, Staughton. 2004. *The Freedom Schools, An Informal History*, https://solidarity-us.org/atc/108/p477/ (accessed February 2019).
Malvern, Benjamin J. Jr. 1966. 'The Purpose of Business: Profit or Service? A Case Report', in Alice Mary Hilton (ed.) *The Evolving Society. First Annual Conference on the Cybercultural Revolution – Cybernetics and Automation*. New York: Institute for Cybercultural Research, 201–208.
Mercel, Jules. 1966. 'Computing Machines: Message and Control Centers of Cybernated Systems', in Alice Mary Hilton (ed.) *The Evolving Society. First Annual Conference on the Cybercultural Revolution – Cybernetics and Automation*. New York: Institute for Cybercultural Research, 35–48.
Michael, Donald N. 1962. *Cybernation: The Silent Conquest. A Report of the Center for the Study of Democratic Institutions*. Santa Barbara, CA: Center for the Study of Democratic Institutions.
Michael, Donald N. 1965. 'Automation. In response to: The Bogey of Automation from the August 26, 1965 issue', *The New York Review of Books*, 25 November, www.nybooks.com/articles/archives/1965/nov/25/automation-1/.
Michael, Donald N. 1966. 'Can Society Adapt Itself Voluntarily and Speedily?' in Alice Mary Hilton (ed.) *The Evolving Society. First Annual Conference on the Cybercultural Revolution – Cybernetics and Automation*. New York: Institute for Cybercultural Research, 209–212.
Michael, Donald N. 1966. Review of *The Age of Automation*, by Sir Leon Bagrit, *Technology and Culture*, 7(2): 257–261.
Mississippi papers. 1964. *International Socialist Review*, 24(3), 85–89.
Nathan, Otto. 1966. 'Our Concept of Economy: Waste or Abundance', in Alice Mary Hilton (ed.) *The Evolving Society. First Annual Conference on the Cybercultural Revolution – Cybernetics and Automation*. New York: Institute for Cybercultural Research, 267–274.
Nieburg, H.L. 1967. 'The Intruders: The Invasion of Privacy by Government and Industry by Edward V. Long', *Technology and Culture*, 8(4): 514–515.
Orlie, Melissa A. 2003. Review of *Turning Operations: Feminism, Arendt, and Politics*, by Mary G. Dietz, *Perspectives on Politics*, 1(4): 760–761.
Pechenkin, Alexander. 2014. *Leonid Isaakovich Mandelstam. Research, Teaching, Life*. Cham/Heidelberg/New York/Dordrecht/London: Springer.
Perks, H.F.W. 1966. 'Economic Aspects of the Evolving Society', in Alice Mary Hilton (ed.) *The Evolving Society. First Annual Conference on the Cybercultural Revolution – Cybernetics and Automation*. New York: Institute for Cybercultural Research, 191–200
Perlo, Victor. 1966. 'Comments from the Floor' in Panel Discussion led by J. T. Schwartz, in Alice Mary Hilton (ed.) *The Evolving Society. First Annual Conference on the Cybercultural Revolution – Cybernetics and Automation*. New York: Institute for Cybercultural Research: 221–250.
Perrucci, Robert and Marc Pilisuk. 1968. *The Triple Revolution: Social Problems in Depth*. Boston: Little, Brown & Co.

Pfeiffer, John E. 1949. 'The Stuff that Dreams Are Made on', *New York Times*, 23 January.
Postel, Danny. 2002. 'The Life and the Mind', *The Chronicle of Higher Education*, 7 June, www.chronicle.com/article/The-Lifethe-Mind/15246.
Ravetz, Jeromy R. 1966. Review of *The Impact of Science on Technology*, edited by Aaron W. Warner, Dean Morse, and Alfred Richner, *Technology and Culture*, 7(2): 255–257.
Riepe, Dale. 1967. Review of *The Evolving Society, Technology and Culture*, 8(4): 524–532.
Robins, Kevin. 1996. 'Cyberspace and the World We Live In', in Jon Dovey (ed.) *Fractal Dreams: New Media in Social Context*. London: Lawrence and Wishart: 1–29.
Schwartz, Leonard S. 1966. 'Communication and the Computing Machine', in Alice Mary Hilton (ed.) *The Evolving Society. First Annual Conference on the Cybercultural Revolution – Cybernetics and Automation*. New York: Institute for Cybercultural Research, 87–92.
Seligman, Ben B. 1966. 'The Social Cost of Cybernation', in Alice Mary Hilton (ed.) *The Evolving Society. First Annual Conference on the Cybercultural Revolution – Cybernetics and Automation*. New York: Institute for Cybercultural Research, 159–166.
Shaw, Tony. 2003. '"Some writers are more equal than others": George Orwell, the State and Cold War Privilege', *Cold War History*, 4(1): 143–170.
Sheehy, Gail. 1964. 'Velvet Voice on Automation', *The Montreal Gazette*, 22 July, news.google.com/newspapers?nid=1946&dat=19640722&id.
Silvey, Ted F. 1966. 'Production Control', in Alice Mary Hilton (ed.) *The Evolving Society. First Annual Conference on the Cybercultural Revolution – Cybernetics and Automation*. New York: Institute for Cybercultural Research, 119–128.
Smith, Thomas M. 1967. Review of *The Challenge of the Computer Utility* by Douglas F. Parkhill, *Technology and Culture*, 8(4): 515–516.
Sterne, Jonathan. 2003. 'Bourdieu, Technique and Technology', *Cultural Studies*, 17(3/4): 367–389.
Sutro, Louis. 1966. 'Evolving Structures of Men and Machines', in Alice Mary Hilton (ed.) *The Evolving Society. First Annual Conference on the Cybercultural Revolution – Cybernetics and Automation*. New York: Institute for Cybercultural Research: 173–190.
The New York Times. 1962. 'Automation Report Sees Vast Job Loss', 29 January.
TIME. 1965. 'The Cybernated Generation' (April).
Wallis, David. 1996. 'After Cyberoverkill Comes Cyberburnout', *New York Times*, 4 August.
Warren, Kenneth. 2003. 'Ralph Ellison and the Problem of Cultural Authority', *boundary*, 30(2): 157–174.
Wiener, Norbert. 1961. *Cybernetics, or Control and Communication in the Animal and Machine*. London: MIT Press.

Wiener, Norbert. 1989. *The Human Use of Human Beings*. London: Free Association Books.
Winthrop, Henry. 1966a. 'The Sociological and Ideological Assumptions Underlying Cybernation', *American Journal of Economics and Sociology*, 25(2): 113–126, www.jstor.org/stable/3485324.
Winthrop, Henry. 1966b. 'Some Roadblocks on the Way to a Cybernated World', *American Journal of Economics and Sociology*, 25(4): 405–414, www.jstor.org/stable/3485779.
Wood, Frederick B. 1964. 'Cybernetics and Public Order', paper presented at the Conference on Cybernetics and Society, Georgetown University, Washington, DC, 19 November, www.fredbernardwood.org/papers/sepr/SEPR93-J-Cybernetics.pdf.
Young-Bruehl, Elisabeth. 2006. *Why Arendt Matters*. New Haven: Yale University Press.

Chapter 5

Polemical acts of rare extremism: Two Cultures and a hat[1]

The Two Cultures debate produced a furore in the modernizing era of the early to mid-1960s. The scientist C.P. Snow's diagnosis of a cleavage that should be healed between the sciences and the arts is still widely invoked. Less well remembered is that his protagonist F.R. Leavis also argued for the benefits of one culture. Not the one arising out of a capitulation to technologically administered utilitarianism, but the culture he discerned within a tradition of community, largely lost in everyday life, but held in the English language and in its literature. Leavis argued that the value of the past in constructing a politics engaging with the constitution of the present should be recognized. It might also inform and found a form of hope for the future.

This chapter engages with Leavis' arguments. The mode of radical liberalism Leavis espoused in the journal *Scrutiny* in the early to mid-20th century produced a response to technology far from technological optimism, but also distanced from Marxist critiques of technocratic rationality. This radicalism is often viewed as hopelessly tarnished by the chauvinistic nationalism that framed and constrained it, which became increasingly marked in later years. However, Francis Mulhern, amongst others, has convincingly argued for a more nuanced reading of Leavis and the 'moment of scrutiny' (Mulhern, 1979), and this prompts a re-reading of Leavis' thinking around the specifically technological and a reappraisal of the position he took at the time of the Two Cultures debate. This is worth doing partly because the combinatory force of an attachment to nation, a distrust of technocratic forms of knowledge and its claims to universality, and a moment of technological expansion has been felt in disturbing ways in recent decades. Brexit's anti-expertise discourse and its appeal to English nationalism, populist movements in the US around Trump, disquiet around algorithmically produced filter bubbles on social media all indicate this. My attempt, to replay the Two Cultures debate from the largely eclipsed Leavis side, is undertaken both to restitute an earlier structure of feeling and to explore the *topoi* that organized it – not least because these are once again attended to closely in our time. What kind of replaying is this? This chapter begins to circulate around the issues at hand via a long-playing record: one that records a performance, that comments on a debate, and that reveals a landscape and a field of cultural contestation once submerged – but now re-arising.

* * *

> Things have come to a pretty underpass in England … (Flanders and Swann, *At the Drop of Another Hat*, 1963)

> Here, then, we have the cultural consequences of the technological revolution. (Leavis, 2013: 86)

Let me play you a record. 'Heat cannot of itself pass from one body to a hotter body … Heat won't pass from a cooler to a hotter … You can try it if you like but you'd far better notter …' Perhaps apocryphally, the largest single audience ever subjected to this disquisition on thermodynamics was the American public on prime-time television, Christmas Day 1967, 'on the occasion of the performance by Flanders and Swann of their catchy ditty on the second law' (Hubbard et al., 1968). The song was from *At The Drop of Another Hat* (1963), a revue show at a London theatre, later an LP. Flanders and Swann were a 'middlebrow,[2] middle of the century, musical comedy act', one of a line of male double acts contributing to a tradition that Andy Medhurst describes as part of the 'national joke' (Medhurst, 2007: 123). Secure, English, with a show that was consciously intimate,[3] it was said of Flanders and Swann that they passed as 'charming amateurs who recreate your living room jocularity without pretending to be professional', whilst being in reality 'urbane and zestful artists … professional down to their fingertips' (Simon, 1967, 105–109). The same theatre reviewer noted claims that the 'pseudo-amateurism' of the Flanders and Swann act was what gave it its characteristic 'British-ness'.

The 'Thermodynamics' song was a pointed take on a dispute over the 'Two Cultures' of art and science that incited England's intellectuals to 'polemical acts of rare extremism' (Mulhern, 1979: 305) fierce enough to produce much national press coverage and popular discussion. Two more Englishmen, C.P. Snow and F.R. Leavis, one a trained scientist, civil servant, and successful novelist, the other a literary scholar, fellow of Downing College, Cambridge, and founder of the journal *Scrutiny*, are yoked together as antagonists when this debate is raised. They are invoked as the personification of cleavage: between the sciences and the humanities, science and art, scientific and literary cultures, and technocratic values versus tradition as desirable logics informing a social order.

The Two Cultures controversy is conventionally understood as the successor to an earlier clash disputing the relative priority of the arts and science; the Huxley and Arnold debate of the 1870s, sparked by Huxley's address at Josiah Mason College in Birmingham (Huxley, 1881: 15,[4] Hultberg, 1997: 196, Stinner, 1989). Huxley argued for the priority of the study of nature over culture, and of science over Arnoldian values (truth and beauty). Nature should lead men, and culture itself, too long locked up by the 'Levites in charge of the ark of culture and monopolists of liberal education', should be transformed to reflect better the new priorities given by 'the definite order' of nature (Huxley, 1881: 3, cited in Hultberg, 1997: 196).

Knocking the Two Cultures debate *out* of its place in this line of succession can be productive. It enables a different kind of auditing of the arguments. Doing this is already to take sides, since Snow cultivated the connection between the earlier interventions and the Two Cultures, whilst Leavis refused it on the disingenuous grounds (given his own splenetic tone) that the earlier debate had been better tempered and conducted between equals. For Guy Ortolano, the obvious disparity between the stakes of an argument around science and the arts in Victorian Birmingham and those of a quarrel in 1960s London and Oxbridge means that this genealogy can be rejected out of hand (Ortolano, 2009, Ortolano in White, 2011: 761–763). Ortolano rather aligns the Two Cultures debate with a series of struggles between 'technocratic' and 'radical' forms of liberalism arising in response to a specific historical constellation (that of post-colonial England in the mid-1960s), one that might help identify some of the 'structures of feeling' (Williams, 1992, 2009) that characterized the (conflicted) sensibilities of that moment. This orientation informs the discussion here. It connects with Francis Mulhern's magisterial account of *Scrutiny*'s history, already invoked, and with his demand that the radicalism of the early Leavis project, beginning in the 1930s with *Scrutiny*, be recognized, even whilst the degree to which it was shattered in later work as internal tensions between materialism and idealism foundational to its project is acknowledged (Mulhern, 1979).

Dislodging the Two Cultures debate from its resting place in a line of arts/science debates makes it easier to also unpick some assumptions about the temporal orientations of the combatants' arguments. Snow's assertion was that the 'Luddite' Leavisites only looked backwards, wishing to dwell entirely in the past. But for *both* sides what was at stake in this argument was not the past but the future, and Leavis' position is more complex and more interesting once this is recognized. The question he sought to address was how to hold a brief for the future whilst also admitting a *preferability* for the past (oddly enough, this emerges as an orientating sensibility in steampunk, gothic, or SF/Cyber noir aesthetics today).

Occlusion

C.P. Snow said the 'Two Cultures' formulation was more than a metaphor but less than a model. It has certainly had an 'enduring afterlife' (Dizikes, 2009),[5] becoming an, if not obligatory, then acknowledged reference or passage point in discussions around the impacts and values of science and technology on society, a synecdoche for 'what is at stake' still invoked in debates around science and culture, technology and culture, hermeneutics and instrumentalism, the stakes of literacy and technical knowledge in an era of computation; arguments about code literacy, hacking versus yacking, programming and education are all relevant here. In a sense, the term invites itself in. It becomes a form of 'currency', in Leavis' (pejorative) sense of the word, offering itself for adoption in relation to emerging formations; one of its recent re-emergences has been in work exploring digital humanities, which latter directly engages with 'English' and its computerization, amongst other things. The persistence of the formulation does not signal full acceptance of what it promotes. It has long been critiqued for the binary division it inscribes; even Snow himself later acknowledged that there might be three or more cultures. Others including Adrian Mackenzie and Andrew Murphie (2008) have argued for multiple, overlapping 'cultures'. But the critiques *are* of Snow's formulation, while the attack by Leavis, constituting one half of the exchange and generating more than half of the controversy at the time, is rarely explored. In public commentary around the fiftieth anniversary of the Two Cultures debates in 2009, it was striking how little was said about Leavis, about the discourse of 'counter-modernity' as he articulated it, or about his critique of unthinking technologization. The focus was on Snow, and the adequacy and relevance or otherwise of his sense of a science/arts cultural cleavage. The debate is remembered in Snow's terms. In this sense he *won*.

Obliteration?

Today Leavis' ideas about technology and culture appear to be largely irrelevant. They have nothing to say to science, whereas, as noted, Snow's *term* still has some bite, even if the arguments the latter introduced under its banner largely do not. In English the Leavisite project was dead by the late decades of the 20th century, eclipsed by structuralism and post-structuralism in various guises, and by the related rise of various forms of feminist and post-colonial thinking inimical to the unexamined universalism of Leavisite humanism. Leavis' considerable contribution to the development of British cultural studies (see later) is now largely disavowed, as Mulhern and others have pointed out. This provides a starting point, since the apparently wholesale deletion of a perspective might be as informing as the perspective itself. It is interesting to explore why (and/or for whom) Leavis' thinking is no longer thinkable at all. Not least because, as the conclusion to this chapter argues, what is no longer 'thinkable' in particular registers, or thinkable as a coherent whole, may continue to inform a sensibility, or to emerge in fragments, to trouble the present. It may be that a mode of hostility towards the rise of technological values that resonates with the Leavisite objections to demands for a particular kind of cultural 'healing' is so thoroughly diffused into everyday discourse that it is hard to tease out and identify, even if it does inform contemporary hostility to new literary formations or approaches.

Raymond Williams[6] characterized as 'residual' those elements of a social structure that relate back to older forms that originally sustained them, now gone, but which remain active (Williams, 2009). The case I am going to make is that Leavis' thinking on technology engaged with, and even contributed to the production of, the intellectual justification for a particular kind of anti-computing, but was already residual, perhaps, even in its time. Today this always already residual mode of anti-computing, which resonates with the Leavisite position, might be extant, and might be a mode whose very amorphousness, as well as its relative impotence against the various claims of 'progress' it seeks to question, renders it tricky to investigate. Williams's sense of the residual, however, even if useful, might underplay, if not the tenacity of earlier forms, then their *capacity* to remap onto new formations and *revive*. Investigating this, at the end of this chapter I briefly consider where Leavis' sense of the relationship between English community and technological 'progress' might be revenant today. And why that is disturbing.

I have no intention of becoming a belated member of the Leavisite 'clerisy'. There is much in the Leavisite programme, including its naturalized muscular, white, view of 'humanity' and its avocation of an increasingly strident nationalism, that repels. Nonetheless, it seems important to examine, rather than simply set aside as unpalatable, or simply not worth re-excavating, the particular kind of hostility to technocratic rationality that Leavis espoused. It constituted a significant strand in a tapestry of responses to early computerization and automation in an England on the edge of cultural change across a series of fronts in mid- to late 20th century. The Two Cultures debate took place near enough to 1963, when, as the poet Philip Larkin famously put it, sexual intercourse was just beginning (Larkin, 1974). Less often noticed, so too was business computing.

The Rede Lecture

Snow's 'Two Cultures' intervention came in the late 1950s, in the Rede Lecture, which was published in *The New Statesman* and in *Encounter*.[7] There Snow identified a rift between the sciences and the arts, claimed that this posed a threat to Britain and/or British influence in a post-colonial era, and called for this rift to be closed through measures resulting in the 'integration of Two Cultures' (Hultberg, 1997: 200), or two 'social orders' – those of literary culture and natural science (Snow 1959: 10). This demanded new forms of literacy and, complaining that literary cultures 'revel ... arrogantly in their ignorance' (Hultberg, 1997), Snow argued that 'basic literacy' (Collini, 1993) should be judged not only by literary criterion but also by knowledge of scientific fundamentals: Shakespeare *and* the First and Second Laws (of thermodynamics) (Snow, 1959: 14). If the text of the lecture is explored in any depth, it is obvious that Snow's real concern was for the *advancement* of science and scientific values *over* other values; as Sam Leith put it, the 'apparent even-handedness of the way Snow articulated the divide is ... deceptive; Snow was taking sides' (Leith, 2009). From Snow's perspective, 'literary intellectuals' hostile to the values of science and ignorant of its basic tenets had come define cultural values in general as 'literary' values. Worse, this kind of valuation was operational in the 'corridors of power' (Snow, 2000), which was *why* it was of consequence. Leith again:

> [Snow] might have regretted all those physicists not having managed to read Dickens but he would not have thought that it actually mattered very much. The scientific illiteracy of the humanities graduates, in Snow's view, mattered very much indeed – and the reason that it mattered was that these were the people in charge of things ... (Leith, 2009).

The (loosely) anthropological register of much of Snow's lecture, at one with the novelistic style adopted in his fictions about public science, and aligning with his interest in science as practice (Hultberg, 1997), might have contributed to acceptance of the egregious 'evenhanded' reading; Dizikes argues that Snow's assumption of the role of 'eagle-eyed anthropologist' dissembled his sense of himself as the 'evangelist of our technological future' (Dizikes, 2009). But the message is clear enough, for Snow, science 'must progress over time' and *will* progress (Snow, 1959: 204), science is the basis of prosperity, security, and 'social hope' (Snow, 1959: 27), and the scientific orientation is progressive and creative, and so, therefore, are those who *embody* this orientation, those who live by scientific values. For Snow, hope (and sometimes 'goodness') inheres in scientists who 'have the future in their bones' (Snow, 1959: 10); as *habitus*, perhaps. Snow's utilitarian or Benthamite argument saw scientific approaches as essential for the common good of the many. His claim was that the social hope science provides could now be made available to human society in general, and specifically to Britain and British people in a world after Empire – these arguments being developed through a discussion of science education and/as British foreign policy. Leith (2009) reads this as a demand to spread the benefits of the scientific revolution to the dispossessed so as to avoid them helping themselves via social revolution. Dizikes, also assessing the global intents of the lecture, describes it as 'irretrievably' a Cold War document (2009).

In contrast to scientists, 'literary intellectuals' are labelled natural 'Luddites', people 'wishing the future did not exist' who can therefore have no relevance in the project of making it (Snow, 1959: 11). Snow indeed comes close to arguing that literary values can *only* be thoroughly retrograde, particularly when various modernist writers whose engagements with fascism were well known at the time are invoked as the, if not typical, then at least *unsurprising* products, of the literary attitude in general – the implication being that literary values *breed* such 'unfree' approaches and attitudes. Close reading of Snow's lecture confirms which group he believes should inherit the future.

The lecture was noticed in interested circles (for instance, provoking letters in the *Encounter*), but did not initially attract general attention. It was Leavis' response, made two years later in the Richmond Lecture (1962), delivered at Downing College and published by the *Spectator*,[8] which launched the debate that became notorious. In his lecture Leavis raged expertly, precisely, and with venom, attacking the argument and the author. Snow, he said, was as 'intellectually as undistinguished as it is possible to be … he doesn't know what he means and doesn't know he doesn't know'. Amongst Leavis' more scabrous remarks was the assertion that Snow's novels must have been written by an electronic brain – perhaps one named Charlie. Only this, he said, could explain the mechanistic writing and lack of life they exhibited (Leavis, 2013: 57). Leavis justified a personal attack on the grounds that Snow had come to embody the ideas he promulgated. He had become 'a portent of our civilization', and attacking him was necessary for forensic reasons. Leavis' claim that this necessity was regrettable wasn't necessarily convincing. Collini comments that:

> A malevolent deity, setting out to design a single figure in whom the largest number of Leavis's deepest antipathies would find themselves embodied, could not have done better than to create Charles Percy Snow. (Collini, 1993: xxxii)

This may be the case. But there is also the proposition that what Snow embodied, and what Leavis attacked, was not only Snow the hubristic scientist and indifferent novelist, but Snow the calculating machine, Snow the computer 'brain' that proceeded by way of mechanistic processing rather than human reasoning. Leavis scathingly attacked Snow's confidence in himself as a 'master-mind', as Collini notes (1993: 9), and it is interesting to consider the degree to which the 'master-mind' in question was devoid of all human qualities. Was it Snow himself, or electronic Charlie, the probability machine, that was the portent?

Bathos

In response to the Richmond Lecture some condemned Leavis' tone but were sympathetic to his argument. *The Guardian* regretted a 'vehement and directly personal' attack on a 'famous social critic' linked with 'the doctrine of the Two Cultures' (*The Guardian*, 1962). The literary critic Lionel Trilling condemned the 'impermissible tone' of the lecture, a remark that circulated widely (Collini, 1993: xxxvii, Matthews, 2004: 60). The violence of Leavis' attack amplified the debate and the resulting 'substantial furore' involved 'most of the print media' (Matthews, 2004: 51) of the day.

The debate had absurd elements. Leavis' raging was notorious, but there was also Snow's pomposity, a bathetic unevenness in register evident in his lecture, badly chosen examples of cultural snobbery and scientific ignorance, and the implication – easy to maliciously infer by Snow's enemies – that he truly believed scientific language should become an everyday language in the drawing rooms of England, or at least in the common rooms of ancient universities. Snow's discussion of literacy is partly framed in relation to his own experiences of being 'brushed off' at the hands of literary intellectuals, and his call for recognition of the scientific as that which is beyond such niceties as cultural capital and beyond individual desires or conceits resolves into a querulous and incongruous demand for due *respect*; 'hear the status symbols/cymbals clash', as Flanders and Swann put it in another song in the same *Hat* show as their 'Thermodynamics' song.

The 'social hope' science offered, identified by Snow as a (post-colonial) global necessity for British foreign policy, also had to do, it turned out, with prestige, power, control, and influence in the English universities. The choice of example and its tone[9] came back to haunt Snow (1959: 14–15) and his supporters (e.g. *The Guardian*, 1962).

Hat performers Flanders and Swann exploited the bathos of the whole affair in their genial but lethal takedown of Snow, which begins with a solicitude as artificial as the domestic setting of the original stage set. The problem that has been raised, say Flanders and Swann, is how to talk to the scientist, who speaks another language – so you must address him in his own tongue: 'ah h2so4 professor ... the reciprocal of Pi to your good wife ... don't synthesize anything I wouldn't synthesize ... this he will understand'. The scientist is to be talked to with the kind of elaborate politeness that replaces casual discourse amongst (true) equals – 'you can't ask him to lend you a quid'. Framing ordinary social intercourse in 'scientific' terms will enable the scientist to take part in the general conversation – or, rather, in what passes for general conversation amongst academic men. The 'Thermodynamics' song, following this banter, completes the attack; 'Heat cannot of itself pass from one body to a hotter body' ... The song is a virtuoso rendition of the First and Second laws and is correct in essentials. It is thus airily implied that *scientific* literacy is easily accomplished – a matter of the right doggerel. Its conclusion goes further, hinting that science might lead down roads that are not worth travelling anyway. It ends with an appeal to the seductive claims of entropy (always appealing to Flanders and Swann, who famously sang of 'mud, glorious mud') over the rigours of constant progress, or constant work; 'heat is work/and work's a curse/and if all the heat in the universe ...'. The project of 'talking to scientists' as well as 'talking science' is thus both accomplished and its absurdity proclaimed. Scientific literacy is delivered and Snow's framing of science/scientists as the bearer(s) of 'social hope', which also came with a demand to be taken seriously personally, is lampooned through a ruthless depiction of the social hopelessness of the scientist. This intervention into the Two Cultures debate, mocking the claims of scientific culture (and the pomposity of academic culture in general), still surfaces in discussions of the art/science divide over the years. More often, however, the song is invoked in accounts of the laws themselves – notably in educational material – but the critique around it is forgotten.

Which England?

The Flanders and Swann Two Cultures skit might appear to have little to say about Leavis specifically, since it is Snow's proposition and his desire for parity in the drawing room that is lampooned. But, playing the record again, a particular kind of context can be discerned arising through a series of numbers skewering early to mid-1960s liberal England and its discontents. These deal in anxieties about technocratic futures, modernity, displacement, the replacement of 'organic community' with new forms of 'civilization' or modern life (Leavis' terms, as discussed below). They include a lament for the land ('Bedstead Men'), a satire on conspicuous consumption ('Design for Living'), on sexual scandal (the Profumo affair), and a pastiche of the public utilities' bureaucratic grip on modern domestic life ('The Gasman Cometh'). What can be audited is a certain hostility to science and technology, one that connects with other values in play, including those around (the fall of) Empire and nation ('The English Are Best'). A satirical paean to the 'triumph' of British engineering of the aeroplane ends when 'the ashtray falls off' – taken as evidence that 'if God had intended us to fly he'd never have given us the railways'. Flanders and Swann also sing of these, but more obviously seriously. Their 'Slow Train' song (1963) celebrates names and places carved into the English landscape, and into the collective memory, by the railway, and mourns the planned closure of many lines under the Beeching axe:

> Millers Dale for Tideswell, Kirby Muxloe ... No more will I go ...
> We won't be meeting again, on the Slow Train ....

In 'Slow Train' an industrial-era technology, having become a settled part of the English landscape, is remarked upon as it passes away.[10] 'The Bedstead Men', meanwhile, comments on the widespread despoliation of the rural countryside. It works through a satirical commentary on the evolution of a traditional form of association; the Bedstead Men litter and dump where their forebears smuggled or robbed highways. They are portrayed, through the words and the forms of the folk song, as absurdly mythic figures, men of the English landscape.

What begins to emerge is a sense that everyday life – its cultures, its *mores*, its cultural memory, and its hope for the future – is somewhat precarious. As Flanders and Swann put it, 'things have come to a pretty underpass in England'. The worldview they offer is of a land experiencing rapid and pervasive transformation. An old way of life is being swallowed up and is sometimes mourned for. The liberalism of Flanders and Swann, given a radical edge by their satirical intervention into political affairs of the day, also has a deeply conservative side. A thick, rich, vein of nostalgia, a commentary on 'progress' and what it leaves behind – old worlds and old social *mores*, ended but not quite replaced – runs through the *Hat* shows. This vein is entangled with the critique of the absurdities of modern life, and of science-speak and its prophets, that more directly comments on the stakes of the Snow/Leavis dispute. Flanders and Swann articulate a worldview not entirely at ease with unexamined progress as it is heralded and delivered through technoscientific advance; least of all when this impacts on traditional forms of everyday life (which they also skewer). Even whilst they deal with the new, and take it on cheerfully enough, their shows offer a more or less liberal commentary on consumerization, technologization, and technocratic rationality in a post-Empire Britain that is also conservative; somewhat resigned – even fatalistic – in tone, and, despite the horseplay, rather serious.

A stage act certainly does not constitute systematic political critique, but this one did make an appeal to a set of (undefined, presumed, assumed) shared sensibilities; those of an imagined community perhaps, one which both performers and audience are presumed to be a part of, even whilst the latter are also systematically mocked. What is articulated is strikingly, and consistently, at odds with Snow's sense that 'social hope' rests with scientific advance, that the latter is the basis for social hope, and is *the* necessary orientation for 'society'.

The anti 'technologico-Benthamite' cultural politics of Leavis, exhibited in the Two Cultures writings and developed over decades, clearly do not align with those of Flanders and Swann (whose metropolitan act would be at odds with *Scrutiny*'s early provincial austerity, apart from anything else). But the themes that emerge in Flanders and Swann's revue act resonate with themes that inform Leavis' body of work, the latter also striking for its advocacy of 'continuity', 'tradition', and 'community' and for its attack on utilitarianism. That's why, in stepping out of the line and going around by way of the *Hat*, it is possible gain a new sense of the arguments informing the Two Cultures debates, and the stakes at the time. Alignments and articulations that have since become naturalized shift and become possible to adjust or disturb. With this sense of realignment in mind I now turn back to explore Leavis' earlier thinking on the technological and to ask how it informs the Two Cultures debate.

Leavis and his sword: the modern crisis

Leavis co-founded *Scrutiny*, the project with which he is most associated, in 1932. *Scrutiny* concerned itself with English literature, finding in English a resource for intervening in culture, understood as literature and as a material form, as language and as life. In *Education and the University* Leavis argued that these terms could not be divided, that there was nothing merely 'literary' about the literary mind (Milner, 2002: 34). The '"governing theme" of *Scrutiny* was industrialization and its destructive effects on society and culture' (Mulhern, 1979: 50, citing Leavis and Thompson, 1933). Decades before the Two Cultures events *Scrutiny* was waging war on technologico-Benthamite tendencies and logics. Industrialization and the automated culture, automated society, and the automated forms of life it produced were regarded as the forces producing a 'crisis of modernity',[11] sometimes adumbrated as the 'modern crisis'. In the *Restatement for Critics* Leavis put this in stark terms:

> is the machine power to triumph or to be triumphed over, to be the dictator or the servant of human ends? (Leavis, 1933: 320 and Leavis, cited in Mulhern, 1979: 60)

Scrutiny writers thought differently about 'machine power'; Leavis focused on how human interventions gave particular characteristics to emerging machines, whilst Denys Thompson's sense was of the fundamental 'spiritual malfeasance' of technology (entailing a version of Innes' bias of technology), for instance. But, albeit by various routes, *Scrutiny* reached a shared position which was, baldly stated, that as a result of industrialization and modernization, itself the product of technology and the structures it produced or enabled, a society of widespread standardization had emerged. This society was characterized by the prioritization of the mass (the 'herd') and of mass decisions, the hollowing-out of bodies and individuals, and the evisceration of language, which was felt to be increasingly losing its capacity to signify (Mulhern, 1979: 55). In sum, there was what Leavis termed an inadequacy of experience, and what Knight, another *Scrutiny* adherent, defined as a sense of a crisis of 'life' itself (Knight, cited in Mulhern, 1979: 75).

Against all this, which is to say against the depredations of industrial commercial 'civilization' and the ascendant values of technological/technocratic rationality, *Scrutiny* asserted the values of continuity, tradition, and above all community, for it was 'organic community' that was under threat. Community was set against 'civilization' and viewed as counterpart to agendas based on the criterion of 'progress', particularly technological progress, both as means and as end. This latter Leavis often termed Wellsian thinking. Community was 'affirmed in tradition' and inhered in social life, though not in social structures. Everyday life was valued because it held a 'residium' of this essential tradition (Knight, cited in Mulhern, 1979: 75), but even this was felt to have become a despoiled source. A 'last sanctuary' could be found, however, in 'the tradition of English literature', which could become a repository of the values *Scrutiny* held dear and might make possible the *continuity* of a form of community. For this reason literary values were held not to be separate from questions of community but, on the contrary, expressed, and should be judged by, standards of life. They are, in a sense, its performance and its archive. Culture (taking literary form in English literature) was thus understood *as* the cumulative meaning of tradition (Mulhern, 1979: 75). Consonant with that, though later, at the time of the Two Cultures debate, Leavis would comment:

> I don't believe in any 'literary values', and you won't find me talking about them; the judgments the literary critic is concerned with are judgments about life ... (Leavis, 2013: 110)

A consequence of this conjunction is that attending to literature may enable a particular kind of space to open up, one that is public or at least *common*. Mulhern aptly sums up *Scrutiny* as embodying a cluster of ideas around the 'nature of literature and its place in social life' (Mulhern, 1979: 328) and as aiming to restore the values of community 'to some kind of authority in the modern world' (Mulhern,1979: 76.). The point is that the Leavisite position did not, in intent at least, signal a *withdrawal*. The English archive was not to become a sequestered, dead thing. *Scrutiny* was a public project and Leavis and his co-thinkers were interventionists. They sought not only to name the modern crisis but to fight for a new form of counter-modernity (see also Mulhern, 1979). *Scrutiny* believed it was right 'to hold to a belief in the preferability of the past, whilst resolving to act in and for the present'. There was to be 'no mere going back', and the engagement with tradition, and specifically the tradition of community, implied more than a simple form of nostalgia (Leavis and Thompson, 1933: 96). In the first issue of *Scrutiny* D.W. Harding argued that it was possible to combine nostalgia with realism, and to distinguish it from regression (see Henstra, 2009: 55). Nostalgia could be radicalized. The memory of the old was to be an 'incitement' towards the new world (Leavis and Thompson, 1933: 97), and there was to be nothing here of an attachment to folk wisdom, always rather despised by Leavis.

Scrutiny ended in 1953. Leavis declared its influence 'decisive' but privately thought it had failed. It had not been embedded in Cambridge English and would, he felt, peter out with his retirement (Mulhern, 1979: 302). However, in his formal response to C.P. Snow, in the Richmond Lecture of 1962 he argued that *Scrutiny*'s attempt to maintain a 'critical function' was relevant to the Two Cultures moment (Leavis, 2013: 76).[12]

Snow's lecture began with art and science, but Leavis began with technology. Holding Snow and his work up for investigation he announced: 'Here, then, we have the cultural consequences of the technological revolution' (Leavis, 2013: 86). For Leavis these consequences were often explored as forms of 'technologico-Benthamism'. The term might be surprising, given the tight association, in the humanities at least, of Benthamism with Foucault's more or less post-human discussion of governmentality and his account of the Panopticon in *Discipline and Punish* (1975; see also During, 1992:

5). Leavis' definition of the term – though not his use of it – is not so very far from Foucault's. For him, technologico-Benthamism was a mode of organizing society that discarded human 'significances, values and non-measurable ends' (Leavis, 1972: 110). It designated those entangled processes of technologically driven systematization, bureaucratization, and individuation that are found at the heart of many resurgent anti-computing anxieties today (when they are often explored in Foucauldian or post-Foucauldian terms). The difference comes in the unqualified defence of humanism that this produces. Amigoni aptly sums up Leavis' attack on Benthamism and the acceleration of the implementation of Benthamite forms of utilitarianism through technology as an attack on 'inhumane' utilitarianism from the perspective of humanism (Amigoni, 2011: 23). Further, if the earlier Leavisite texts indicate that this general orientation was long standing, the Richmond Lecture indicates that it was constantly being revived in relation to new technologies – and particularly new communicational and computational technologies emerging in the 1960s.[13]

Inhumanity – set against full humanity – was key in Leavis' response to the contention that literary intellectuals were backwards looking. First, he made a distinction between marking a loss and acknowledging its consequences and simply repining. Community of a particular kind, based on craft, or rural life, for instance was not 'something we should aim at recovering; but … something finally gone' (Leavis, 2013: 107), on the other hand, marking this loss did matter – and industrialization, Leavis argued, had entailed heavy losses. He attacked the contention that it had involved a simple decision, a freely made and painless choice, by a previously agricultural society, to leave the land, as Snow had implied. This description, he felt, failed to acknowledge the full complexity of the process, as a historical process, and above all failed to frame it as a 'full human'[14] problem (Leavis, 2013: 70). It also produced a viewpoint within which no proper understanding of the contemporary position could be had or would be countenanced:

> if you insist on the need for any other kind of concern, entailing forethought, action and provision, about the human future – any other kind of misgiving – than that which talks in terms of productivity, material standards of living, hygienic and technological progress, then you are a Luddite. (Leavis, 2013: 64)

Pasts and futures

Leavis' argument turned not only on what was gone but needed to be marked, but also on what remained live. These were the stakes in setting (his) 'cultural tradition' against (Snow's) adumbration of 'traditional culture'. Cultural tradition was *live*, demanding not reposeful mourning, but wakefulness. It could traverse the community, and had done so until, threatened by industrialization in general, and technologico-Benthamism or technologically driven and delivered modes of new utilitarianism in particular, it was pushed back to a last stronghold in English literature. But even in those conditions, argued Leavis, continuity – the old watchword of *Scrutiny* – pertained. For Leavis there is a continuity *between* the material culture of an organic community and its literary articulation. So, whilst Snow identified the presence in contemporary society of two cultures, openly demanded their integration, and covertly lobbied for the triumph of the scientific orientation, as the dominant form, over that of the literary, the burden of Leavis' argument was (also but very differently) that there could only be one culture (Leavis, 2013).

For Leavis, continuity is grounds for hope and is the means through which he will come to the 'explicit positive note that has all along been my goal (for I am not a Luddite)'. The cultural tradition he valorizes allows for questions of the future to be raised because it is a continuous living tradition, based in a present reality, and lived by humans 'on the spot', in that place where they find themselves (Leavis, 2013: 67), that place in which they may recognize themselves perhaps, because they are, or to the extent that they are, embedded in tradition. To be on the spot is thus to be in tradition and to be in the present. Leavis contrasts this with Snow's offer for a society built on the promise of 'jam tomorrow' which 'enjoins us to do our living in the dimension of "social hope"' (Leavis, 2013: 69). For Leavis, 'jam tomorrow', or the promised benefits of technologically driven expansion and plenty, was what came of relying not on cultural tradition but on the promises of growth driven by technology. He claimed that his objection was not to this kind of program per se, but to the one-sided reliance on the forms of knowledge it would provide and principles it would work through:

> To point out these things is not to be a Luddite. It is to insist on the truth that, in an age of revolutionary and constantly advancing

technology, the sustained collaborative devotion of directed energy and directing intelligence that is science needs to be accompanied by another, and quite different, devotion of purpose and energy, another sustained collaborative effort of creative intelligence. (Leavis, 2013: 108)

Leavis' response to Snow, his argument that tradition and culture be protected from the assault of the technocrats and their 'Benthamite' evisceration and standardization, was thus both an analysis and a demand. True to his earlier principles, Leavis set continuity and tradition against the specializing, divisive, eviscerating, amnesiac qualities of a world reorganized according to the principles of a new form of technocratic rationality, which he understood Snow to embody. This was his rejoinder to Snow's demand for specialist cultures.

Leavis' attack on social hope was integral to his rebuttal of Snow's (quasi-anthropologically based) argument that scientists themselves form a culture *because* they, unlike non-scientists, 'think alike', or *are* alike. In his lecture Snow called scientists 'creative rather than critical', adding that as a consequence of this they tended to be 'good-natured and brash' (Hultberg, 1997, 202 citing Snow). His cheerful scientists, whistling their way to the neutron bomb, or in his novels seeing no way out of this trajectory except betrayal (see the *Strangers and Brothers* series, and in particular *Corridors of Power*, 2000 [1964]) stand against the mordant English traditionalists, and the two camps already look very different. However, 'thinking alike' in Snow's hands also entails a form of individual renunciation. The point for Snow is that 'social hope' will be delivered by the specialist general community of science for whom *personal* wishes and desires are irrelevant and selfish (they are perhaps 'good-natured' by training). The principles that make for best practice for scientists are based on impartial calculation of best-case scenarios for populations in general, generating work and setting priorities in favour of the 'greater good', defined as that which brings progress to humanity. Literary intellectuals meanwhile are set down as antisocial, culpably selfish for taking their own desires into account in bemoaning industrialization and its acceleration, and as living in the wrong tense.

By contrast, for Leavis the individual is the locus of decision, of compass, and of judgement. He can thus ask: What is the 'social condition' that has nothing to do with the 'individual condition'? What is the 'social hope' that transcends, cancels or makes indifferent the inescapable tragic condition of each individual (Leavis, 2013: 65)? There is culture and its traditions, bound into community, and the individual who constitutes it. Without individuals, and this is at the heart of the bodily attack on Snow and the attack on the body of his work, there can be no human values but only a civilization based on technocratic values, and quantifiable ends, which deals with units of population. For Leavis, the computer–writer, the C.P. Snow automaton, is the personification, or perhaps better the *instantiation*, of this technologico-Benthamite orientation (and in, this sense, is already not human at all), and threatens the very existence of the last repository of culture and continuity – since how can 'English' be defended without human readers? Leavis' anti-computing quip, his comment that the writer of the Rede Lecture was quite possibly a computer, is thus deadly in intent. The computer-writer Charles Snow *replaces* the human with the computer and obliterates sense and judgement, founded in the human individuals and confirmed through that durable engagement between them that finds value or significance, and might constitute a community. In the place of this is put a moment of calculation. Significance is replaced by statistical 'solutions'. As Leavis had put it long before, in *Nor Shall My Sword*, the result of such thinking would be that 'we need take no ends into account in our planning and calculating but those which are looked after by quantitative criteria, the statistical: "quality", that is will take care of itself' (Leavis, 1972: 138).

If the protagonists in the Two Cultures dispute were contending for 'humanity' and who 'has' it, then a peculiar irony here is that attack on the 'computer' Charles who/which threatens to usurp 'life' is made in the context of the *Leavisite* substitution of English literature – and the guardian of its canons – for those forms of 'life' that previously constituted 'organic' community, that were held in other forms of culture. The material that stands against the computer and its code is the text and its language; English literature, which for Leavis has become the repository of what would once have been part of a general sensibility (one that was not, to use the term 'dissociated', as T.S. Eliot had put it in the early 1920s). Arguably, there are no 'full human' bodies left on *either* side of this dispute.

Leavis' general defence, as it was constructed, found resonance with many. Whether or not his insistence that his version of cultural tradition was not simply nostalgic was convincing, it appeared to connect with a structure of feeling in England at that time – something like that found in some of the Flanders and Swann songs invoked earlier; those combining a sense of change and loss, of something moving away at a fast pace, with a sense of the rise of new absurdities emerging as a consequence of technological modernity. Leavis' sense was that control was in danger of being lost, ceded to machines and machine makers. This was now a matter not (only) of abandoning culture to look for laws in nature (these were Huxley's terms, it will be recalled) but also of inventing the mechanisms to recapitulate these laws on social grounds (Snow on science and the social good), and he argued that it gave away something important: human autonomy and maturity. This too found support; *The Spectator* mounted a cautious defence of Leavis, or at least a rebuttal of unthought-through technological progress, in terms that refer to this sense of infantilization:

> If we think of some purpose then we shall be able to make the future instead of being carried along by it like children in the arms of automation. (*The Spectator*, 1962)

Non-inevitable consequences

Others, by the time of the Two Cultures debate, already had 'purposes' and visions of possible futures that were not necessarily those of *The Spectator*, or Snow, or Leavis. With childhood and adulthood invoked, we could say these others sought to grow up differently. To conclude, then, let me briefly here return a final time to the *Hat* performances, which took place more or less in that space Larkin adumbrated – somewhere after the Chatterley trial and before the ascent of the Beatles,[15] a space that held possibilities that were, as Larkin noted, somewhat inaccessible to those native to an older time.

In the final verses of the 'Thermodynamics' song, heat and work (energy and its transformations) are translated back to 'culture'; 'heat is work' (says physics) and 'work's a curse' (said emerging social movements of the time) – and 'all the heat in the universe' had 'better cool down' (say Flanders and Swann). Here is a sudden glimpse of a different future; an era of 'cool'. This is only a glimpse and it is laid aside.[16] In the end, the finale (offered by science, provided in song) is entropy, a universal brownness[17] covers all – in glorious mud. Flanders and Swann thus publicly give up on both kinds of cool and go back to the drawing room, their furbelowed retreat. They never intended to be hip and fashionable, nor to espouse shiny new science. Their act says that they know their place, but it also signals that they know their *time*. They were (and I suppose always are, as often as the LP is spun up again), on the cusp of being superseded by a rising youth culture and its counter-cultural influences. Neither they, nor their performances, were avant-garde, as they acknowledged off-stage in various places.

All the old men and England too

Leavis and Snow, that other pair, were also of a generation that was, at the time of the Two Cultures debate (and the *Hat* shows) *being moved* (in) on. The form of science and the literary formations they espoused were both changing. On the one side there was big science (not so called at the time) and the nuclear issues, and the UK's changing role in the post-war world, which Snow got at only very imperfectly (and from a perspective that is now difficult to understand). And on the other side a change in relative values of various literary formations, including the emergence of powerful new forces within them such as structuralism/post-structuralism with its challenge to such English and 'English' values as 'universal' humanism, and, eventually, to humanism itself. Leavis and Snow were arguing about divisions and boundaries, and about a set of associated values, and rules for setting values, that were *all already* changing rapidly; the ground was moving under their feet. As Matthews points out, neither Leavis' high style nor the ethnographic realism Snow adopted in his literary endeavours found connections with, for instance, the Angry Young Men, whose work had already energized the cultural scenes of the 1950s. Matthews claims that the 'importance of the literary critical field' was 'already waning' by that mid-1960s, and that this waning was 'caught up in the Two Cultures debate' (Matthews, 2004: 62); for him it is partly *why* the debate happened. Others too, have noted that the Two Cultures debate was one occurring within the establishment, that it took place whilst (other) events and other groups, less parochial than

Snow (with his limited sense of globality) and more cosmopolitan than Leavis (with his insistent sense of *English* culture) were emerging. These events were challenging the definitions of 'science', and the boundaries of 'community', and the relationship between community and nation, and the normative constitution of 'the individual' – scientist or not.

Snow and Leavis argued about the primacy of particular views of the world; about the right or even the capacity of the scientist or the literary intellectual to hold the future in his hand. And here, as a pointer to a series of exclusions that need to be *marked*, we might turn to gender – and return once again to the peculiarly personalized terms of the old debate. That is, we might ask *who* is the science professor of C.P. Snow's lament? What do we know of 'him', in 'human' terms? Only three things: that he speaks 'science' ('this he will understand'), that he is a man with a wife ('the reciprocal of pi to your good wife'), that he wishes to enter the common room on equal terms with those others in the arts who can lay claim to it, perhaps those who (may) have wives to talk of. Those, then, who are – normatively/overwhelmingly – men, not women.

Perhaps this seems like a diversion. Let me divert further. In a review of Brenda Maddox's book on DNA scientist Rosamund Franklin and her work with Watson and Crick, Sarah Delamont considers the question of how to assess Franklin and her work on DNA in relation to gender issues. She notes Maddox's claim that Watson thought about calling his double helix book *Honest Jim* – after Kingsley Amis' novel title, *Lucky Jim*. Maddox also suggests that Watson's destructive 'caricature' of Franklin in his book is very like that of Margaret Peel in *Lucky Jim*. The academic on whom Peel was based (Larkin's lover) was also caricatured in Bradbury's *Eating People Is Wrong*, which is *also* a Flanders and Swann song line. Catch a breath. As Delamont notes, such stereotypes (and I only gesture here to how Delamont's crawl might be extended still further) constitute a 'salutary reminder of how deep misogyny ran in British universities in the 1950s, in both of C.P. Snow's Two Cultures' (Delamont, 2003: 315). This was always a conversation amongst men. Two public school boys on stage, and two old grammar boys in different common rooms, even if one of them claimed he was always an outsider, constituted two sides of a shared and in many ways exclusive world.

The scandal of the Profumo affair (of which Flanders and Swann sang obliquely in 'Horoscope' and 'Madeira'), and the changing role of 'sexual intercourse' in public culture in England, might be used to mark the beginnings of a shift. When somewhat later the Two Cultures began to be explored or contested in relation to the building of a different kind of 'science as culture' it was feminists who were often central to this inquiry; and feminist theory that was influential. The question of automation and control – of what it might mean, of how we might respond, if we *were* children, or women, or new kinds of hybrids or cybrids finding ourselves 'in the arms of automation', could be, and was, posed in very different ways only shortly after the Two Cultures debate. Notably the project to develop new kinds of (British) cultural studies, formally set up at Birmingham with a grant of £2,400 from Penguin Books in 1963, and a remit to cover everything from 'Leavis to the *News of the World*' (*Guardian*, 1963), came to disavow largely its Leavisite roots and to cleave to a left rather than a liberal agenda. It also became a hub for much work on feminism, and for much feminist work – but only when cultural studies itself had been, as Stuart Hall put it much later – broken into.

Conclusions

Taking seriously the Leavisite side of these debates, exploring the possibilities arising out of radical nostalgia, and attempting to understand or acknowledge the pain of change in a time of transformation, partly using Flanders and Swann to link into a minor key sensibility of the period, might provide a restitutive reading of Leavis' arguments in the Two Cultures; the decision to attack Snow personally appears in a different light as it becomes oddly impersonal. However, both sides of the debate threaten to resolve into each other; Leavis' fervour and Snow's brashness merge. Snow's demand for change is really a demand for continuity, and the radicalism in Leavis' demand for the future as human (the ultimate continuity) is lost because his sense of the force of the past and its capacity to intervene in the future is undermined by the limitations of his universalist (but in fact chauvinistically local) sense of human culture;

the insistence on English, always a struggle to comprehend, in the end becomes a sign of parochialism rather than a radical attack on elites that seeks to undermine their privilege and their insistent demands for unexamined 'progress' at any cost.

However, if Leavis' thinking is unfashionable within most academic spheres and is virtually ignored in relation to questions of technology as these are related to digital cultures, it does nonetheless find echoes today. The anxieties Leavis articulated persist and frequently return, resonating with new forms of 21st-century modes of technological hostility. Where would the Leavisite objection to technologico-Benthamism and its indifferent standardization, the global extension of a nonetheless nationally chauvinistic position, and the nostalgically powered defence of a folk and a land find its resonances today? Not only as a sensibility but as the re-constitution of a more active formation in response to that? One immediate answer in an England in post-Brexit referendum times is in new forms of national politics – perhaps, and sometimes despite the intentions of those holding such positions, a politics with its attractions for those of the new Right.

Notes

1 'inciting England's intellectuals to polemical acts of rare extremism' (Mulhern, 1979: 305).
2 See Raymond Williams (2017) on middlebrow.
3 Their set included curtains, a piano, and a domestic lightshade – if not a common room, then its domestic counterpart, the drawing room.
4 'Nature is the expression of a definite order with which nothing interferes, and ... the chief business of mankind is to learn that order and govern themselves accordingly ... this scientific "criticism of life" presents itself to us with different credentials from any other. It appeals not to authority, nor to what anybody might have thought or said, but to nature' (Huxley, 1881: 15).
5 'The two cultures', a phrase with an 'enduring afterlife' (Peter Dizikes, *New York Times*, 19 March 19th, 2009).
6 Williams himself has been termed a 'left Leavisite' and *New Left Review* was intended to fuse Leavisite literary criticism with radical left politics (see Milner, 2002: 25).

7   Also printed in *The New Statesman and Nation* in 1956.
8   *The Guardian* commented somewhat snidely that it was 'almost the only 'posh' paper not attacked by the doctor in the lecture' (Yesteryears of Snow), and also noted that *The Spectator* was published in the heart of Bloomsbury.
9   Snow had delivered his lecture in a style that alternated between self-preening earnestness and airy high-table bonhomie. The position of the Rede lecturer, however, is one that is implicitly congratulatory (Leith, 2009).
10 Leavis also commented on railways in 'One Culture', citing Dickens (Leavis, 2013: 95).
11 'modernity' – tin cans and computers.
12 'I am thinking again of what Scrutiny stood – and stands – for: of the creative work it did on the contemporary intellectual-cultural frontier in maintaining the critical function.'
13 A few years after the Richmond Lecture Leavis would respond to Harold Wilson's 'white heat' speech and the agenda it promoted.
14 'If one points out that the actual history has been, with significance for one's apprehension of the full human problem, incomparably and poignantly more complex than that, Snow dismisses one as a "natural Luddite"' (Leavis, 2013: 70).
15 George Martin produced the *Hat* LPs.
16 Via the obscure interjection 'Oh, Beatles – nothing!'
17 'And universal Darkness cover all' (Alexander Pope, *The Dunciad*).

Bibliography

Amigoni, David. 2011. *Victorian Literature*. Edinburgh: Edinburgh University Press.

Baxter, Alan G. 2000. 'Michael Flanders, Karl Landsteiner and a belief in fairies', Broadcast 25 June, www.jcu.edu.au/cgc/Flanders.html.

Boughton, J.M. 2002. 'The Bretton Woods Proposal: A Brief Look', *Political Science Quarterly*, 42(6): 564.

Burnett, D. Graham. 1999. 'A View from the Bridge: The Two Cultures Debate, Its Legacy, and the History of Science', *Daedalus*, 128(2): 193–128.

Cohen, Benjamin R. 2001. 'Uniquely Structured? Debating Concepts of Science, from the Two Cultures to the. Science Wars', Thesis, Virginia Tech University.

Collini, Stefan. 1993. 'Science: Two cultures: Still Natural Enemies?', *The Independent*, 15 November, www.independent.co.uk/news/science/science-two-cultures-still-natural-enemies-stefan-collini-examines-how-far-c-p-snows-idea-of-an-1504486.html.

Collini, Stefan. 1998. 'The Critic as Journalist: Leavis after Scrutiny', in J. Treglown and B. Bennett (eds) *Grub Street and the Ivory Tower*. Oxford: Oxford University Press, 150–176.
Collini, Stefan. 2013. 'Introduction' in F.R. Leavis, *Two Cultures? The Significance of C.P. Snow*. Cambridge: Canto Classics.
Collini, Stefan. 2015. 'Whisky out of Teacups', *London Review of Books*, 19 February.
De la Mothe, John. 1992. *C.P. Snow and the Struggle for Modernity*. Texas: Austin University Press.
Delamont, Sarah. 2003. Review: *Rosalind Franklin and Lucky Jim: Misogyny in the Two Cultures, Social Studies of Science*, 33(2): 315–322.
Dizikes, Peter. 2009. 'Sunday Book Review', *New York Times*, 19 March.
During, Simon. 1992. *Foucault and Literature*. London: Psychology Press.
Flanders, Donald and Swann, Michael. 1963 *At the Drop of Another Hat*. EMI 7243 8 29399 2 4.
Flanders and Swann. 1963. *The First and Second Law*. Flanders and Swan online, www.nyanko.pwp.blueyonder.co.uk (accessed 13 December 2011).
Foucault, Michel. 1975. *Discipline and Punish*. London: Penguin.
Fraknoi, Andrew. 2007. 'The Music of the Spheres in Education: Using Astronomically Inspired Music', *The Astronomy Education Review*, 1(5): 39–153.
Harding, D.W. 1932. 'A Note on Nostalgia', *Scrutiny*, 1(1): 8–19.
Henstra, Sarah. 2009. *The Counter-Memorial Impulse in Twentieth-Century English Fiction*. Basingstoke: Palgrave Macmillan.
Hubbard, W.N., P.A.G. O'Hare and H.M. Feder 1968. 'Experimental Inorganic Thermochemistry', *Annual Review of Physical Chemistry*, 19: 111–128.
Hultberg, John. 1997. 'The Two Cultures Revisited', *Science Communication*, 18(3), 194–215.
Huxley, Thomas. 1881. *Science and Culture and Other Essays*. London: Macmillan.
Larkin, Philip. 1967. 'Annus Mirabilis', in *High Windows* (1974). London: Faber & Faber.
Leavis. F.R. 1933. 'Restatement for Critics', *Scrutiny*, 1(4): 315–323, www.unz.org/Pub/Scrutiny-1933mar-00315.
Leavis, F.R. 1953. 'The Responsible Critic, or the Function of Criticism at Any Time', *Scrutiny*, 19(3): 162–183.
Leavis, F.R. 1962. 'The Richmond Lecture', delivered at Downing College, published in the *Spectator*, 9 March. Produced in book form by Chatto and Windus later that year.
Leavis, F.R. 1967. *English Literature in Our Time and the University: The Clark Lectures*. London: Chatto and Windus.
Leavis, F.R. 1968. *A Selection from Scrutiny*. Cambridge: Cambridge University Press.
Leavis, F.R. 1972. *Nor Shall My Sword: Discourses on Pluralism, Compassion and Social Hope*. London: Chatto and Windus.

Leavis, F.R. 2013. *Two Cultures? The Significance of C.P. Snow*. Cambridge: Cambridge University Press.
Leavis, F.R. and Denys Thompson. 1933. *Culture and Environment: The Training of Critical Awareness*. London: Chatto and Windus.
Leavis, Q.D. 1984. *Collected Essays: Volume 1. The Englishness of the English Novel*. Cambridge: Cambridge University Press.
Leith, Sam. 2009. 'The Split between Scientists and Writers', *Financial Times*, 9 May, www.ft.com/intl/cms/s/0/3a2dfeb6-3b5e-11de-ba91-00144feabdc0.html.
Mackenzie, Adrian and Andrew Murphie. 2008. 'The Two Cultures Become Multiple?', *Australian Feminist Studies*, 23(55): 87–100.
MacKillop, Ian Duncan. 1995. *F.R. Leavis: A Life in Criticism*. London: Allen Lane.
Matthews, Sean. 2004. 'The Responsibilities of Dissent: F.R. Leavis after Scrutiny', *Literature and History* (3rd series), 13(2): 49–66.
Medhurst, Andy. 2007. *A National Joke: Popular Comedy and English National Identity*. London: Routledge.
Milner, Andrew. 2002. *Re-imagining Cultural Studies: The Promise of Cultural Materialism*. London: Sage.
Mulhern, Francis. 1979. *The Moment of Scrutiny*. London: New Left Books.
Mulhern, Francis. 2013. 'English Reading', in Homi Bhabha (ed.) *Nation and Narration*. London: Taylor and Francis, 250–264.
Ortolano, Guy. 2009. *The Two Cultures Controversy: Science, Literature and Cultural Politics in Postwar Britain*. Cambridge: Cambridge University Press.
Samson, C. 1980. 'Problems of Information Studies in History' in S. Stone (ed.) *Humanities Information Research*. Sheffield: CRUS: 44–68.
Simon, John. 1967. Theatre Chronicle/Arts Review, in *The Hudson Review*, 20(1): 105–114.
Snow, C.P. 1959. *The Two Cultures and the Scientific Revolution*. The Rede Lecture (May), Cambridge.
Snow, C.P. 2000. *Corridors of Power*. Cornwall: Stratus.
Snow, C.P. 2012. *The Two Cultures*. London: Canto Classics.
Stinner, Andrew. 1989. 'Science, Humanities and Society – the Snow–Leavis Controversy', *Interchange*, 20(2): 16–23.
Tasker, John. 1972. *The Richmond Lecture, Its Purpose and Achievement*. Swansea: Brynmill Publishing.
The Guardian. 1962. 'The Significance of C.P. Snow', 20 March (unsigned article).
The Guardian. 1963. 'A Study of "Pop" Culture', 7 June (unsigned article).
The Spectator. 1962. 'The Two Cultures' (Editorial), 29 March.
Treglown, Jeremy. 1998. 'The *TLS* in the Second World War and How to Fill Some Gaps in Modern British Cultural History' in J. Treglown and Bridget Bennett (eds) *Grub Street and the Ivory Tower*. Oxford: Oxford University Press, 135–150. (p. 136).

Ward, Philip M. 2000. 'Science and the Arts: Two Cultures or One?' A talk for National Science Week, House of Commons (March), www.pemward.co.uk/page_1158331496750.html.
Wellek, René. 1981. 'Reviewed Works: F.R. Leavis: Judgment and the Discipline of Thought by Robert Boyers; The Literary Criticism of F.R. Leavis by R.P. Bilan; The Moment of 'Scrutiny' by Francis Mulhern', in *The Modern Language Review*, 76(1): 175–180.
Whelan, Robert. 2009. 'Fifty Years on, C.P. Snow's "Two Cultures" are united in desperation', *Daily Telegraph*, 5 May.
White, William. 2011. Review of 'The Two Cultures Controversy: Science, Literature and Cultural Politics in Postwar Britain by Guy Ortolano', *English Historical Review*, CXXVI: 761–763.
Williams, Raymond. 1981. *Culture*. London: Fontana.
Williams, Raymond. 1992. *The Long Revolution*. London: Hogarth Press.
Williams, Raymond. 2009. *Marxism and Literature*. Oxford: Oxford University Press.
Williams, Raymond. 2017. *Culture and Society*. London: Penguin Vintage Classics. [First published 1958]

Chapter 6

Apostasy in the temple of technology: ELIZA the more than mechanical therapist

> What is it about the computer that has brought the view of man as a machine to a new level of plausibility? (Weizenbaum, 1976: 1–16)

> We, all of us, have made the world too much into a computer, and … this remaking of the world in the image of the computer started long before there were any electronic computers. (Weizenbaum, 1976: ix)

What happens when technophilia falls out of love with its object? In the early 1970s, Joseph Weizenbaum, a computer scientist at the Massachusetts Institute of Technology (MIT), an institution 'proudly polarized' around technology, wrote a script for a 'computer program with which one could "converse" in English' (Weizenbaum, 1976: 371). This program, strictly speaking a language analyser and script,[1] became famous as a chatbot (Wardrip-Fruin and Montfort, 2003) and was once labelled 'the most widely quoted computer program in history' (Turkle, 2005: 39). ELIZA, whose name came from Shaw's heroine in *Pygmalion*, carries out natural language conversations with the user, inviting interaction through the 'impersonation' of a Rogerian therapist (Wardrip-Fruin, 2009, Weizenbaum, 1976: [2]). Strictly speaking, ELIZA was ELIZA playing DOCTOR (the script for a therapist), but it was by that name that the program became celebrated.

Weizenbaum chose to give ELIZA this identity because of the kind of interaction that Rogerian therapy entails, involving extensive forms of mirroring of the patient by the therapist. Essentially it 'draws out the patient' by 'reflecting back' (Weizenbaum, 1976: 3). This kind of dialogue is peculiarly amenable to computer simulation because it is open to wide input but demands a limited repertoire of responses (as output) and has a clear context (Weizenbaum, 1972 and 1976,

Turkle, 2005). The kinds of responses that Eliza could produce might seem less tendentious or eccentric in these circumstances.

Weizenbaum always recognized that ELIZA had many limitations, being easily 'persuaded' to come out with clearly nonsensical answers or fall into recursive loops. ELIZA operated rather badly, he said, doing far less than was popularly claimed – which was, at its most exaggerated, to have demonstrated a 'general solution to the problem of computer understanding of natural language', a key AI goal (Weizenbaum, 1976: 7). These limitations did not prevent ELIZA from becoming a phenomenon, generating discussion, hype, and debate inside and outside specialist and public worlds. For Weizenbaum this reception was egregious and damaging. It provoked a permanent and radical shift in his thinking around the relationship between humans and computers specifically, and humans and technology more generally. As a direct response, he wrote *Computer Power and Human Reason*, an attack on the growing reliance on, and preference for, computational thinking in popular and specialist circles. This was anti-computational thinking from a man at the heart of things, and constituted apostasy in the temple. It was a direct attack on the 'hubristic' arguments of the fast-growing AI community, or the artificial intelligentsia as Weizenbaum termed them. He critiqued in particular the adherence of members of this community to 'the deepest and most grandiose fantasy that motivates work on artificial intelligence ... to build a machine on the model of a man ... which can ultimately ... contemplate the whole domain of human thought' (Weizenbaum, 1976: 203–204).

This chapter begins by considering Weizenbaum's critique of what he understood as the prioritization of computational thinking over human reason. First, the roots of Weizenbaum's change of heart and the development of the arguments in *Computer Power* are explored. Second, I attend more closely to what ELIZA offered, which was, after all, and despite Weizenbaum's protestations, somehow a mode of 'therapy'. Finally, I suggest, partly as a thought experiment, that ELIZA, the famous bot psychiatrist, is itself *conflicted*, dealing in narratives and yet operating in code. Was ELIZA a *computer* with anti-computational impulses? If so, then Weizenbaum might have taken solace from his own creation, even if he remained horrified by the rise of machine thinking in general. At least in that, viewed this way, ELIZA was on his side.

The argument here is in dialogue with, and relates closely to, a parallel article exploring ELIZA in *AI and Society* (Bassett, 2019),[3] which traces these ideas into the contemporary moment via an engagement with human 'droning' (Andrejevic, 2015) by exploring the connections between this early form of automatic 'therapy' and the modulation of selves through behaviouristic forms of 'therapeutic' nudge. Here I stay closer to the Weizenbaum/ELIZA events as they unrolled at the time. I am interested in the workings of apostasy; the abandonment of a belief or principle, traditionally one that pertained to religion. Being true to what you know, if that implies a kind of camp loyalty as well as a form of expertise, turned out to be impossible for Weizenbaum. On the other hand being true to its 'nature' was, or so I am going to argue, a more complex issue in the case of ELIZA than might at first appear. And ELIZA, at least as much as Weizenbaum, is the subject of this chapter. So, here goes.

In *Computer Power and Human Reason* Weizenbaum listed three things he found shocking about ELIZA's celebrity reception. The first was that ordinary people related to ELIZA as if it was a 'real' therapist. The second was that ELIZA was given a warm reception from some in the world of professional therapy. The third was that what ELIZA could do technically was misunderstood and exaggerated. Weizenbaum always maintained that ELIZA's capacity to act as a therapeutic mirror was extremely limited and that the role was chosen 'for convenience', as an aid to researching natural language processing and not with the aim of developing working artificial therapy tools. However limited it was, the program held up a mirror reflecting expert and public understandings of, and hopes for, technology. What Weizenbaum saw reflected there provoked him to speak out.

Computer Power and Human Reason was a best-seller, a narrative of conversion that attempts to convert in its turn. It is at once, and avowedly, a personal account, an expert intervention, a mode of public science, even a science literacy project. In it Weizenbaum binds his thinking into philosophical critiques of technocratic rationality but also names himself as part of another tradition. He invokes Mumford, Arendt, Ellul, and Roszak, amongst others, as key critical thinkers of the technological with whom some of his positions intersect, but speaks as a scientist rather than a philosopher. Indeed, it is *because* he is a scientist that he feels he *needs* to speak, and *may* speak with some authority.[4] He thus finds an appropriate forebear in the Soviet scientist Polyani, who began as a physical chemist but later became chiefly a lifelong critic of 'the mechanical conception of man'; Polyani's response to the Soviet leader Bukharin's contention that science for its own sake would disappear 'under socialism' was that (Soviet) socialism's 'scientific outlook' was in danger of producing 'a mechanical conception of man and history in which there was no place for science itself ... nor ... any grounds for claiming freedom of thought' (cited in Weizenbaum, 1976: 1). Weizenbaum's thinking ran along similar lines, as we will see, but there was another resonance too; Polyani's decision to respond to Bukharin's provocation led to a lifelong change of direction, demanding 'his entire attention', even if he had originally intended to 'have done with it in short order' (Weizenbaum, 1976: 2), and even at the time of writing *Computer Power* Weizenbaum felt himself to be reliving this trajectory. By the second decade of the 21st century his reputation rests far more on his critique of computation than it does on his substantial contribution to computer science and natural language processing.

Computer Power attacked forms of 'powerful delusional thinking' about computers, their capacities, their limits, and their possible impacts, said to be then circulating in expert and public realms. The concern was that they lent credence to, and might accelerate, a broader form of scientism that was already very pronounced. This attack is two pronged; a critique of various ways in which computers are being thought *about* is undertaken and computer thought *itself* is reconsidered. At the heart of the matter, in both cases, is the contention that human thought and AI cannot be equated, so that one may not entirely substitute for the other. This translates into an inquiry into 'whether or not human thought is entirely computable' or reducible to logical formalism, the latter characterizing AI and/or generally taken to constitute its ontology.[5] Weizenbaum's argument was that computer logics/logical formalism will always be 'alien' to human forms of rationality and that the failure to recognize fundamental differences between human rationality and computational reasoning produces errors in (technical) understanding, and a series of social pathologies. Moreover, so his argument went, this error was *already* widespread in his time, being evidenced in over-optimistic expectations for AI, in behaviourism as a tool for social order and a way of understanding humans; in the spread, in other words, of systems thinking and those 'mechanical conceptions' of man that, earlier, Polyani too had feared. More proximately, it was evidenced in the responses to ELIZA, which demonstrated the currency of this kind of thinking, as well as threatening to contribute to it.

Logical formalism?

'First there is a difference between man and machine.' It was Weizenbaum's understanding of what machine logic *is* that led him to argue for an essential difference between humans and machines, for recognition of the division between rational thought and logical operations, and, following on from that, to make a call for limits to the spheres of operation in which the application of machine logics should go ahead. Key to this was an attack on what Weizenbaum termed the 'rationality-logicality' equation. This is the tendency to presume that human rationality and its operations can be reduced to, or can be equated with, or treated as, a question of logic and its operations – *and that this can be operationalized*. For Weizenbaum it cannot. There is an ontological distinction between human intelligence (and human rationality) and computer logic, and because of that also a difference between computer operations which may entail forms of emergence (today that issue arises in terms of machine learning) and human becoming.

Weizenbaum's argument doesn't rely on the outcome, successful or otherwise, of various tests based on simulation (the Turing test being the most obvious example). In this he more or less follows Searle (1980), who is invoked in *Computer Power*. Nor is the issue purely a matter of the possible technical limits to AI development, or certainly not as these were invoked by Dreyfus around the same time (Dreyfus, cited in Weizenbaum, 1976: 12). Nor is it relevant whether or not AI could become complex enough to produce an adequately convincing simulation of humans; for Weizenbaum, '[e]ven if computers could imitate man in every respect – which in fact they cannot – even then it would be appropriate ... urgent ... to examine the computer in the light of man's perennial need to find his place in the world' (Weizenbaum, 1976: 12).

Weizenbaum did accept that imitation approaches were likely to progress, that computers would come to 'internalize ever more complex and more faithful models of ever larger slices of reality' (Weizenbaum, 1972: 175), but believed this process would hit limits. His argument entailed first, a general take on computer processing as a mode of formalization, and second, a consideration of processing itself as a mode of formalization that always entails a form of abstraction and *therefore* a reduction; there is always a gap between what is represented, or symbolized (formalized or encoded) and the subject of such operations – for instance, the human or human thought processes. This understanding set Weizenbaum against AI's basic tenets, as the psychologist Sherry Turkle noted in her 1980s consideration of 'second selves', pointing out that AI has historically asserted 'the primacy of the program'. The AI method, that is, 'follows from the assumption that what looks intuitive can be formalized, and that if you discover the right formalization you can get a machine to do it'. It was a belief in the generalizability of the program, or system (cybernetics is clearly relevant here), that allowed AI to proclaim itself, 'as psychoanalysis and Marxism had done, as a new way of understanding almost everything'[6] (Turkle, 2005: 245–246).

Simulation criterion as a test for intelligence (or a test for successful formalization) follows from this worldview, and, as Turkle noted, AI communities tended to accept the simulation criterion (Turkle, 2005: 267). Simulation as a test for intelligence was rejected by Weizenbaum because he didn't believe in the rationality-logicality equation (what we might define as the generalizability of the AI formalization of intelligence). In particular, he didn't believe human rationality could be systematized, certainly not as, or through, 'logicality' or forms of machinic systematization. As he put it, 'hardly any' of man is determined by 'a small set of formal problems' but, rather, by his (*sic*) intelligence. The consequence is that 'every other intelligence, however great, must necessarily be alien to the human domain' (Turkle, 2005: 223).

Turkle explicitly links Weizenbaum's sense of the 'un-codable' nature of the human to Searle's famous Chinese room argument, the end point of which is that a computer simulation of thought is not thought. In simulation, intelligent machines 'will only be simulating thought' (Weizenbaum, 1976: 263, Searle, 1980). What is being contested here is partly that line of thinking that says the program is not only a way to understand how to *model* human behaviour but how to account for what it is. Or does. If simulation is an acceptable criterion it is because if it operates successfully; if the program runs, if it *plays*, then that is enough. For Weizenbaum, however, the model (perhaps any model) is not the reality. Moreover, and specifically in relation to the modelling of the human, Weizenbaum followed the line that said, given the particular and singular character of human intelligence, it may *never* be possible to convincingly model that reality. His critique of abstraction in general was thus supplemented by his sense that the sum of human knowledge (or knowledge about what the human is) could never be encoded in information structures. Why? Because 'there are things human beings know by virtue of having a human body, or as a consequence of having been treated as human beings by other human beings' (Weizenbaum, 1976: 209). Moreover, even those things that seem communicable through language alone, not appearing to rely on embodiment, necessarily *invoke* it – even if in a distanced way.

Weizenbaum, surprisingly perhaps, invoked Shannon's information model, noting that even here the expectation of the receiver and their history are relevant to what is communicated. Translating this into the register of communication and memory, he commented that 'the human use of language manifests human memory … giving rise to hopes and fears', but then shifts the focus from memory and the past into a comment on the place of the future in human thought. His argument essentially was that, across time, in relation to each other, through their embodied state, and by virtue of these relations/locations, humans are always 'in a state of becoming' and always in a state of becoming which rests on humanity, on the human 'seeing himself, and … being seen by other human beings, as a human being' (Weizenbaum, 1976: 223). It is human becoming, which constitutes human rationality, that cannot be reduced to machine learning. For Weizenbaum, it would be 'hard to see what it could mean to say a computer hopes' (Weizenbaum, 1976: 207).

Weizenbaum did not deny that the advent of AI, defined as a 'subdiscipline of computer science' (Weizenbaum, 1976: 8), had changed human–machine relations, nor that it might redefine human-to-human relationships and/or human relationships or interactions with the world. The adoption of a medium-focused model to explore this is striking. What is described is a particularly tight 'coupling', a binding[7] to the machine, a specific mode of prosthesis operating differently from earlier forms of human–technology interaction, engaging directly (human) 'intellectual, cognitive, and emotive functions' (Weizenbaum, 1976: 9). This coupling is partly why the computer is such a powerful new instrument and why the specific question of limits arises as a proximate necessity. Weizenbaum's answer to the question of what it means to be a human being and what it means to be a computer is thus that these two forms of being, and the forms of intelligence they bear, are fundamentally distinct. One state of being cannot be rendered into another. The trouble begins first when computers and humans move close to one another, or are *bound* to one another in new ways, and when it is assumed, perhaps partly as a consequence of this relation, that this binding should become increasingly tight. His second concern was that this binding up does not produce an exchange, or make a new common ground, but produces a shift towards formalization, or the adoption of computer or machine models as ways to understand the world and/or to make it *operational*.

Machine metaphors

Weizenbaum, then, was concerned with accelerating computer power but also saw limits to the goals of the 'artificial intelligentsia': this is not an account fearful of the 'rise of the machines' to become our new overlords. Weizenbaum was afraid of what he had produced and wanted to cut it down to size *despite* knowing it to be technically limited, and incapable of much that was claimed for it. His concern was that humans were already living *as if* the promises of AI were real, *already* taking their logics for granted. He feared a world in which humans 'come to yield … [their] … autonomy to a world *viewed* as machine' (my italics). Something about the computer he knew had given this view 'a new level of plausibility'(Weizenbaum, 1976: 8), and after ELIZA he urgently wanted to understand what that *something* was; the more urgently because its effects were being felt not only in computer science and AI fields, but also amongst therapists and the general public.

If Weizenbaum was horrified by the apparent alacrity with which a bad model was accepted, and an unfortunate metaphor (his description of ELIZA's designation 'therapist') taken up, it was also because he felt this betrayed a *wish* to believe, a *wish* to lean on computational intelligence, a lack of humans' confidence in their own being, an increasing belief in the superiority of other (scientific, machinic, exact) forms of reasoning, calculation, determination, and an apparent desire to rewrite the social world to reflect these forms. The seeming ease with which the replacement of people with computers could be contemplated troubled him, producing urgent questions of the relationship between the individual and the computer and 'the proper place of computers in the social order'.

The trouble began with computer science and AI where the rationality-logicality equation was being operationalized and promoted. If rationality is understood in terms of logicality, then the proposition that human thought is computable becomes easy to accept. Weizenbaum's objections to this position, strikingly close to some currently recapitulated in relation to big data, led him to consider a broader shift, the diminution or termination of the human's role in 'giving meaning to his world', that function now being taken over by the computing machine (Weizenbaum, 1976: 9). The logico-rationality 'equation' would not produce an *equal* exchange but a prioritization of one set of values over the other. Computational, or, speaking simply, 'scientific' thinking, *prioritizes* logical operations and the knowledge they produce. Weizenbaum noted the degree to which young computer science scholars had 'rejected all ways but the scientific to come to know the world', but was more worried that this orientation had diffused still further and was widely observable. Science, whose values he has championed, in whose temple he still lives,[8] had, he said, become a 'slow acting poison' (Weizenbaum, 1976: 1–16),[9] challenging and transforming human activities and values across a wide range of fields because it offers a particular understanding of humans and social worlds as machines:

> the attribution of certainty to scientific knowledge by the common wisdom, an attribution now made so nearly universally that it has become a commonsense dogma, has virtually delegitimized all other ways of understanding. (Weizenbaum, 1976: xx)

Part of Weizenbaum's fear was that this rendered governmentality inhuman. Social and human contradictions and antagonisms become 'merely apparent contradiction[s]', to be viewed as problems that 'could be untangled by judicious application of cold logic derived from some higher standpoint'. If rational argumentation is really only 'logicality', which follows if rationality itself has been 'tragically twisted so as to *equate* it with logicality' (my italics), then real human conflicts are to be viewed as simply 'failures of communication' to be sorted by 'information handling techniques'. More fundamentally, if there are no 'human values' that are incommensurate, that are *not* sortable by machine, then 'the existence of human values themselves' is in doubt (Weizenbaum, 1976: 13–14).

Mechanical humans and 'automatic good'

Now to return to ELIZA and the reception accorded to this most notorious script. How could so much be hung on such a flimsy piece of work? What could a piece of work like ELIZA do that mattered? My sense is that ELIZA's pertinence came about at least partly because of that something Weizenbaum always wished to downplay, indeed bitterly regretted having assisted in producing; he gave ELIZA to the world as a therapist. And the program was often received as such. Reflecting back on ELIZA, Sherry Turkle notes that 'ELIZA was a dumb program', but adds that it was one that 'sophisticated users' could relate to 'as though it did understand, as though it were a person' (Turkle, 2005: 40, 39).[10] Weizenbaum had earlier noted with exasperation that well-educated people, even computer science specialists and those in circles that might be expected (he felt) to know better, found something compelling, and personal, about their interactions with ELIZA. Notoriously, his secretary asked him to leave the room whilst she 'talked' to ELIZA. Weizenbaum understood this partly in terms of a misplaced anthropomorphism (Weizenbaum, 1976: 6) taking the form of what might be termed reciprocal imprinting; having no better model, humans seeking to understand machine intelligence tended to draw on the only model of intelligence to hand – their own.

The transposition human–machine also worked the other way around – when it meant understanding humans in machine terms.

This disturbed Weizenbaum still more, particularly when he discerned it circulating amongst people who claimed expertise in understanding what makes humans, rather than machines, 'tick' – to invoke a comparison he would presumably have hated. Thus, in the opening pages of *Computer Power* Weizenbaum invoked the psychiatrists who 'believed' the DOCTOR computer program could grow into a 'nearly complete' automatic form of psychotherapy (Weizenbaum, 1976: 3), incredulous that this group, of all people, could 'view the simplest mechanical parody of a single interviewing technique as having captured anything of the essence of a human encounter'. He concluded that it was possible only because they must *already* think of themselves as 'information processor[s]', adding that therapy *did* already have a mechanical conception of the human to hand in the behaviourist theories of B.F. Skinner (Weizenbaum, 1972: 175, 1976). It also had an informational model of the human to draw on in developing AI and computational science research, notably at MIT itself, where Marvin Minsky had defined the human as a 'meat machine' (see also Weizenbaum, 1972: 160).

Weizenbaum's resistance to automatic therapy plugs in to fierce debates within therapeutic circles that were circulating at the time of ELIZA's launch. Central here was the rise of behaviourism, notably as espoused by Skinner. Skinnerian behaviourism is noted for its rejection of an inner self as motivating human actions and its focus on genetic endowment and environment. The key to psychological change in this form of intervention is conditioning via environmental modification, with the aim being the production of positive feedback loops that generate new forms of good behaviour (Skinner, 1971, Bassett, 2019). Behaviourism thus reduces the human to an element within a system that may be stimulated and adjusted as necessary to produce the desired/desirable outcomes. Notoriously, Skinner came to define this as 'automatic goodness', elaborating the term in *Beyond Freedom and Dignity* (Skinner, 1971), published within three or four years of *Human Reason*, and linking it explicitly to forms of governance. *Time* magazine summed up its message as: 'familiar to followers of Skinner, but startling to the uninitiated: we can no longer afford freedom, and so it must be replaced with control over man, his conduct, and his culture' (*Time*, 1971).

Behaviourism's refusal of agency, its denial of the self, its desire to delegate matters of decision around good and evil to agencies beyond the individual human, was anathema to many. It was challenged by left theorists, including Noam Chomsky,[11] for its societal implications and for the morality of its desire to replace freedom with conditioning as a form of social control (Chomsky, 1971a, 1971b, Bassett, 2019).

Another notable critic was Carl Rogers, whose eponymous approach to therapy at the time was focused on self-actualization, stressing the growth of the self and self-autonomy. Rogerian therapy was person centred. In contrast to behaviour-modification programmes designed to change actions in the external world, it explored projects of work on the self, to be undertaken by the subject with self-actualization as the goal (Rogers, 2004). The latter was defined by Rogers as:

> ... the curative force in psychotherapy ... man's tendency to actualize himself, to become his potentialities ... to express and activate all the capacities of the organism. (Rogers, 1954: 251)

Skinner and Rogers later debated their positions in public (see e.g. *A Dialogue on Education and Control*, Skinner and Rogers, 1976), with Skinner defending the virtues of 'automatic goodness' and demanding recognition of its necessity in a complex society, where freedom could not be afforded. Rogers was horrified by the implications of understanding human motivations and actions in machinic ways, as only matters of response to stimuli, and by what followed – which was that therapy, or more broadly the therapeutic modulation of human behaviour to make a better society, would then best consist of the appropriate modulation of conditions to encourage correct forms of behaviour for individuals and also groups. He insisted on the need for thinking through matters of orientation, decision, agency and freedom. Rogerian therapy does not turn the human into an object, thereby diminishing their agency, but on the contrary seeks to allow growth.

So what kind of therapist was ELIZA? One the one hand, a language analyser and script; a piece of code operating to parse speech and offer appropriate responses, a bot designed to host interactions, where formally speaking meaning – understanding – has not been designed in. To process natural language does not entail *understanding* it – not then, not now. On the other hand, ELIZA had been designated Rogerian, and not simply designated, but also *designed* to operate, as a Rogerian. If the program had a goal it was to offer such therapy (or simulate such a therapeutic response) or, in other words, to help 'users' self-actualize, to understand themselves better, to know their own stories. Code and machine, database and narrative: I have argued elsewhere that the one and the other are not discrete, and also that one *survives* the rise of the other (Bassett, 2007). Here we might say that ELIZA provided or enabled *both* for her users. Or perhaps we can say He/She/It was conflicted; not the status a therapist is meant to admit to but, rather, to diagnose (but then, don't all therapists have their own supervisors?). Is this simply a fantasy? Weizenbaum said ELIZA's designation wasn't intended to be taken seriously, the burden of his argument being, perhaps, that the program was designed to be illustrative (of how a natural language processing issue could be addressed in relation to a human function), rather than performative (acting out a role and producing what it helped to name).

It could be concluded, then, that ELIZA *simulated* a (Rogerian-style) therapist rather too successfully for Weizenbaum's liking. But it is also possible to conclude that an assemblage that included ELIZA (providing one form of cognitive input) and human users did, in *operation*, enable forms of reflection or exchange that might produce insight or growth, and that these exchanges need to be considered as neither entirely 'alien' nor wholly (if wholly implies exclusively) 'human'. ELIZA was made of code, language, an analyser and a parser, and in this sense could offer only a machinic view of the world, but what the program was programmed to do, or 'be', was Rogerian; it was programmed not to modulate but to 'listen', to reflect back, and prompt introspection, self-inspection, self-growth.

When Turkle talked to doctoral students at MIT about ELIZA she concluded that they liked talking to a machine, partly because they weren't happy with humans. But another conclusion might be that these students engaged with an interlocutor to develop their thinking/selves. This is far from Skinnerian pigeon boxes, cybernetic loops involving the adjustment of behaviour through the delivery of external stimuli, routing entirely around the sense of self. Put this another way: what did those people, in front of the mirror, do? They told ELIZA stories about themselves. ELIZA invites narrative.

If ELIZA is a narrative machine, then we might suggest that the program is in conflict with what its ontology seems to suggest; there is some hope here that because we are humans we can make of machines – and of machine prosthesis, that magical bind – something, some form of engagement, that is new. Human becoming is not machine emergence, but the two together might be something else again.

This invites reconsideration of what it means to be anti-computing, or even what it means that what we make with computers is often apparently in tension with what is conventionally assumed to be essential to the computational: its inescapable logico-rationality. The latter is in evidence, but also in evidence is an imaginary, a symbolic, a form of putting into practice, or becoming through mediation, which is a coming into the world. The computational as instantiated, *because* instantiated, contains something of its own contradictions. What this opens the door to is a way of rethinking anti-computing not in terms of appropriate use and appropriate limits (where Weizenbaum ends), but perhaps rather as something integral, something reflecting the difficulties, or ambiguities of a complex prosthesis, a prosthetic arrangement, which may still be worth undertaking. To consider the ways in which a machine might conflict with its own ontology – how it might be anti-computing, considered in its own terms, not as a rational choice, but through how its logics are operationalized – comes close to identifying a form of essential indeterminacy, in process if not in use.

The conclusion of *Computer Power and Human Reason* is a call for recognition of a fundamental difference between alien and human thinking, and a call for limits to 'what computers ought to be put to do' (Weizenbaum, 1972: 16). This comes as a demand not for a formal limit on the development of particular computer capacities but, rather, for a restriction on the areas in which they might be used. This makes sense, since for Weizenbaum, in the end, 'the relevant issues are neither technological, nor even mathematical but ethical. If these issues are not addressed', he concludes gloomily, it is because we have finally or already 'abdicated to technology the very duty to formulate questions' (Weizenbaum, 1976: 611).

It would be possible to argue that Weizenbaum had only himself to blame for the ELIZA events. He had programmed his script to imitate a therapist, to mimic, that is, precisely that mode of interaction that cleaves very closely to questions of 'the human', of 'human being', and of human thoughts, feelings, mind. Weizenbaum wrote of the 'perverse proposition' that a computer could be programmed to become an effective psychotherapist. (Weizenbaum, 1976: 206), but this was what he had 'asked' ELIZA to do; in a sense, it was his proposition. His apostasy perhaps arises partly because he felt responsible for creating a (small) monster – a familiar-enough sentiment found amongst scientists who mess with 'life', and one explored extensively in fiction, from Frankenstein (and his parent) on. To break with Weizenbaum I have, rather than excoriating ELIZA as an unintentionally proffered accelerant for accelerationism/machine rationality whose influence came from her attributed capacities rather than anything substantial, reconsidered what ELIZA did – and specifically what ELIZA did *as* a therapist.

An afterword

Plug & Play, a documentary directed by Jens Schanze, made many years after the ELIZA events, intercuts interviews with Weizenbaum with comments from Ray Kurzweil, the singularity champion, and Hiroshi Ishiguro of the Bits and Atoms lab at MIT, still the temple of AI[12] and the now the home of Ishiguro's peculiar robot doppelganger. In *Plug & Play* Weizenbaum affirmed his belief in human inexactitude, change, finitude, and argued passionately against the charms of living forever. He died during the making of the film, his life ending, or so we understand, somewhere near where he began it in Europe, as a Jewish kid – *which is to say in history as well as in technology*. The film ends with an empty chair, and with the silence of Weizenbaum's machines. In his absence they have nothing to say and are – disturbingly – not lively. Dead flesh and unliving machines; there is a difference. ELIZA 'lives' on the internet. You can visit her, have an audience in front of her Rogerian mirror, and wonder perhaps how she relates to Siri, Alexa, and her other later and more commercially minded and linked-in 'sisters'.

Notes

1 ELIZA is a 'language analyser' and 'script'. The latter is described by Weizenbaum as a set of rules 'rather like those that might be given

to an actor who is to use them to improvise around a certain theme' (Weizenbaum, 1976: 3).
2 www.cs.umd.edu/hcil/muiseum/weizenbaum/joseph_page.htm
3 Elsewhere I have looked at further at the return to behaviourism, which haunts this book, and haunts our current situation; from Facebook prompts to behavioural economics to or 'hyper-nudge' (see Yeung 2012) – an increasingly automated affair. If we were drawing a line from the ELIZA moment today, and followed Weizenbaum, we might see in contemporary developments the consummation of a particular trajectory; a greater loss of self, and a massively accelerated adoption of machine organization of individual and social behaviour. I've addressed these strains elsewhere (Bassett, 2019).
4 Weizenbaum is generally remembered as the expert, the insider, who changed sides. An example is found in Wardrip-Fruin and Montfort's (2003) collection, where his thinking is distinguished from 'dire warnings' often come from 'non-computing users' who have only superficially considered or even experienced new media.' Lewis Mumford, for the humanists, acknowledged that it sometimes matters that a member of the scientific establishment says some things that humanists have been 'shouting about' (quoted in Agassi, 1976).
5 See Beatrice Fazi (2018) for an elegant demand to reconsider computational determination.
6 Turkle (2005: 246) footnotes that she uses the term AI to refer to 'a wide range of computational processes, in no way limited to serial programs written in standard programming languages'.
7 This is oddly magical: a binding spell.
8 The contexts of Weizenbaum's cogitations were painfully evident to him. He worked and wrote at MIT and declared himself 'professionally trained only in computer science, which is to say (in all seriousness) that I am extremely poorly educated' (Weizenbaum, 1976: 371).
9 On the genealogy of accounts of technology as cure or poison see also Plato, Derrida, Stiegler.
10 Turkle claims that many people first became aware of hackers through Weizenbaum's description of the hollow-eyed young men reminiscent of opium addicts and compulsive gamblers found at MIT.
11 Noam Chomsky's 'The Case Against B.F. Skinner', a review of *Beyond Freedom and Dignity* that amounted to a coruscating attack on the social programme of Skinnerism, set out to demolish its scientific rigour and denounce the morality of conditioning as a mode of social control (Chomsky, 1971b; see also Bassett, 2019).
12 Where the future was 'invented', according to Stuart Brand's (1998 [1987]) eponymous book.

Bibliography

Agassi, Joseph. 1976. 'Review of Computer Power and Human Reason: From Judgment to Calculation', *Technology and Culture*, 17(4): 813–816.
Andrejevic, Mark. 2015. 'The Droning of Experience', *Fiberculture Journal*, 25: 187.
Bassett, Caroline. 2007. *The Arc and the Machine*. London: Manchester University Press.
Bassett, Caroline. 2019. 'The Computational Therapeutic: Exploring Weizenbaum's ELIZA as a History of the Present', *AI and Society*, 34: 803–812, https://doi.org/10.1007/s00146-018-0825-9.
Brand, Stuart. 1998 [1987]. *Inventing the Future at MIT*. London: Penguin.
Chomsky, Noam. 1971a. 'Skinner's Utopia: Panacea, or Path to Hell?' *Time*, 20 September.
Chomsky, Noam. 1971b. 'The Case Against BF Skinner', *The New York Review of Books*, 30 December.
Fazi, M. Beatrice. 2018. *Contingent Computation: Abstraction, Experience, and Indeterminacy in Computational Aesthetics*. London: Rowman & Littlefield International.
Freedman, David. 2012. 'The Perfected Self', *The Atlantic*, June, www.theatlantic.com/magazine/archive.
Geller, Leonard. 1982. 'The Failure of Self Actualization Theory', *Journal of Humanistic Psychology*, 22(2): 56–73.
Kurzweil, Ray. 1999. *The Age of Spiritual Machines*. London: Viking.
Kurzweil, Ray. 2005. *The Singularity Is Near: When Humans Transcend Technology*. London: Duckworth.
Rogers, Carl R. 1954. 'Toward a Theory of Creativity', *ETC: A Review of General Semantics* 11(4): 249–260.
Rogers, Carl. 2004. *On Becoming a Person*. Constable: London.
Searle, John. 1980. 'Minds, Brains and Programs', *Behavioral and Brain Sciences*, 3(3): 417–457.
Skinner, B.F. 1971. *Beyond Freedom and Dignity*. New York: Knopf.
Skinner, B.F. and Carl Rogers. 1976. *A Dialogue on Education and Control*, www.youtube.com/watch?v=olg4Az_hV4Y (accessed 16 May 2021).
Stiegler, Bernard. 2013. *What Makes Life Worth Living: On Pharmacology*. Oxford: Polity.
Sunstein, Cass R. and Richard H. Thaler. 2008. *Nudge: Improving Decisions about Health, Wealth, and Happiness*. London: Penguin.
Time Magazine. 1971. 'B.F. Skinner says we can't afford freedom'. *Time* cover, 20 September.
Turkle, Sherry. 2005. *The Second Self: Computers and the Human Spirit*. London: MIT Press.
Wardrip-Fruin, Noah and Nick Montfort. 2003. 'Introduction to Weizenbaum', in *The New Media Reader*. London: MIT Press, 367–368.
Wardrip-Fruin, Noah. 2009. *Third Person: Authoring and Exploring Vast Narratives*. London: MIT Press.

Weizenbaum, Joseph. 1966. 'ELIZA – a computer program for the study of natural language communication between man and machine', *Communications of the ACM* 9(1): 36–45.
Weizenbaum, Joseph. 1972. 'How Does One Insult a Machine?' *Science*, 176: 609–614.
Weizenbaum, Joseph. 1976. *Computer Power and Human Reason, from Judgment to Calculation.* San Francisco: W.H. Freeman.
Will, George F. 2008. 'Nudge against the Fudge'. *Newsweek*, 21 June, www.newsweek.com/george-f-will-nudge-against-the fudge-90735.
Yeung, Karen. 2012. 'Nudge as Fudge', review article, *Modern Law Review*, 75(1): 122–148.

Filmography

Plug & Pray. 2010. Jens Schanze (director).

Chapter 7

Those in love with quantum filth: science fiction, singularity, and the flesh

In this chapter questions about AI that ELIZA foregrounded are explored in new places and times – in science fiction, which has long dealt in AI, singularity, and the computational. SF claims a privileged relationship to the technological future, and the tax on dissenting projections is lower than that for the apostates of computer science and industry. More specifically, it claims the privilege that comes with attention. It attends to the future, it explores, invents, and/or speculates on possible forms of life. Through the form of attention SF pays, it creates, it makes, and that which is made in fiction is made possible through the fictional, the speculative, the fantastical. The philosopher Paul Ricoeur argued that narrative may resolve the aporias of time by making these aporias productive on another order of language. It is plausible to claim for SF, then, that it produces real (possible) futures in poesis. It is in this way also producing accounts of the present. Recognizing the tendency of utopian and dystopian accounts to reverse their charge, this chapter avoids polemical accounts of AI 'life' as good or evil and explores aspects of the anti-computational in more ambiguous explorations of fictional future being.

* * *

> I think hard times are coming when we will be wanting the voices of writers who can see alternatives to how we live now and can see through our fear-stricken society and its obsessive technologies. We will need writers who can remember freedom. Poets, visionaries – the realists of a larger reality. (Le Guin, 2014)[1]

A widespread popular (and also literary) presumption is (still) that SF is enchanted with technology, dazzled by its shiny promise, its solid materiality, and its capacity for inflicting lethal damage at new scales. For this reason it is often castigated for exhibiting an unexamined technophilia, and/or for adopting crude forms of technological determinism (the judgement pertaining to utopian and dystopian SF). The opposite view was promulgated by the novelist Ursula Le Guin, who near the end of her life declared that SF writers were important precisely because they can see *beyond* the enclosures of technocratic rationality and its temporal horizons. Warning against what she saw as the 'obsessive technologies' of a 'fear-stricken society', Le Guin argued that SF (she refused to divide 'speculative' and 'science' fiction) constitutes a response to an extant mode of technocratic domination. SF practitioners can gain an expert view of the present *because* as the 'realists of a larger reality' they can generate a complexity around thinking the technological that is lacking in a world in which technology is obsessively lauded or relentlessly excoriated. For Le Guin this is why SF can develop real alternatives to 'how we live now' – integral to which is what life itself is or might become in the future. It is life and its fictional treatment that is the subject of this chapter. The focus is on singularity, defined as that expected or feared moment when technological advancements mean that humanity augments either to the point of becoming something qualitatively new or is superseded and left behind by the rise of new forms of artificial consciousness. SF has long had dealings with the tensions and paradoxes at the heart of singularity discourse, with its demands to upgrade, augment, arise, upload, to create and/or terminate new and old forms of human and other life.

In this chapter, aspects of changing life are explored in Marge Piercy's *He/She/It* (known in the UK as *Body of Glass*), China Miéville's *Embassytown*, and Hannu Rajaniemi's *Fractal Prince* series. Between them they span almost twenty years, from early cyberspace and the first AI revival to today's much larger re-emergence. These bodies of work provide insights into how post-standard-human life has been viewed as possible, viable, or desirable, and in what forms, at significant moments in the decades-long processes of computerization beginning around the early public internet/PC era and coming into the near present. In various ways they all exhibit something of the kind of 'greater realism' Le Guin advocates. This places them at some distance from the claims of singularity advocates, which lean for their authority on scientific discourse, and often on what Kate O'Riordan terms 'unreal objects' (O'Riordan, 2017); speculative future technologies pulled into the present, as if they were already fully functional, or *as near as makes no difference*. What is made for real in SF, which declares and rests upon its fictionality, it thus allowed to contrast with what is not yet made 'for real' but yet claims to be non-fictional in discourses circulating around singularity science. Further, all the works invoked challenge the position that the technologically given augmentation of human or machine intelligence is an automatic good, expressing ambivalence or dislike for, or suggesting the preferability of, particular forms of life. I would argue that bound up with SF's interest in, or excitement around, technology as an instrument of change there is always an impulse that is more ambivalent or even hostile, that hostility as much as avidity is part of its genre identity, an unacknowledged or sequestered part of the broader range – and certainly not confined to fully dystopian SF.

Singularity

Singularity marks a tipping point. It is defined as that moment to come when the rise of AI means that what it is to be human changes qualitatively. That is, we humans are changed, and/or our position in the world changes as a consequence of the rise of new kinds of intelligences that out-smart us. Singularity stands for the inauguration of that time in which, through the advancement of technology, humans at once become more (because augmented) and less (since further from any baseline – or normative – model) than human. It also refers to the break that occurs as AIs take on more or less conscious (or certainly highly agential) lives of their own. A long-standing argument within transhumanism concerns the relative merits of continued embodiment, albeit in augmented form, versus various forms of uploading in which the human and/or flesh body is left behind entirely (upload fantasies began early in singularity discourse, notably in Ray Kurzweil's writing (2005)). A singularity moment would constitute an extreme transduction. It breaks out of any predictable temporal progression, not only being a matter of scaling up. Max Tegmark explores singularity as Life 3.0 and says it comes about when 'hardware dependence' on the human body to support intelligent life is ended (Tegmark, 2018). Another advocate for singularity-delivered transhumanism, Russell Blackford, concludes it will bring about 'deeply altered people ... continuous with us, but unlike us in many ways'. He adds that 'optimistically', these new people will *be* 'us ... greatly changed'; pessimistically, they are humanity's *successors* 'a new race, the inheritors of the earth ... usurpers' (Blackford, 2013: 442). For hard singularity advocates the question is not *that* singularity will come, but what happens *when* it does: augmentation or new life, expanded human or fully artificial being? A key marker of singularity discourse is its reliance on big science and big industry, on digital, biotechnological, robotics, neuroscience enterprises to deliver on this future. This alignment has consequences; it is not only the crude determinism of singularity science as it is invoked by singularity advocates or believers that jars with critically orientated theorizations of emerging techno-cultural change, but often also its market-driven, neoliberal, or libertarian political orientations.

Critical and cognitive theory and singularity

Critical theory tends to distance itself from singularity science discourses, particularly in their populist forms, whilst itself entertaining a series of somewhat discrete positions. In an article in *Existenz* Francesca Ferrando (2013) helpfully disambiguates by dividing transhumanist thinking on singularity from forms of critical post-humanism, which begin by decentring technology as the determining cause and then further dividing post-humanism approaches. Notably, anti-humanism is defined directly against transhumanism because it encapsulates approaches to thinking singularity that reach beyond simple technical questions of augmentation/inauguration and its discontents/contents. Metahumanism, defined through the invocation of del Val and Sorgner's 'A Metahumanist Manifesto' (2011), is different again. Drawing on forms of more or less Deleuzian[2] entanglement, it is against the prioritization of the shaping limits of the body (that which defines 'the human') in thinking about intelligent life, and against the invocation of *human* ideas of free will, autonomy, and rationality as distinctions that imply the superiority of the anthropoid over future forms of life. The meta-human body is thus regarded as post-anatomical, as marking a 'development away from humanism'. Such a body cannot be individual. Thus a 'common relational body' is postulated; this body is liberated from particular kinds of bodily constraints (or disciplining), but it is also always in danger of (re)appropriation (del Val and Sorgner, 2011).

The burden of Ferrando's argument is that critical post-humanism, particularly the metahuman variety, is better at grappling with emerging questions of human and post-human future existence than transhumanism in any of its political 'flavours' (libertarian and/or democratic transhumanism, or extropianism), transhumanism being compromised both by the degree to which it leans on science and technology and by the concealed ultra-humanism, or 'fit for all' idea of the human, which it contains. These two issues, which are conjoined, severely restrict transhumanism's perspectives, rendering it unable to grapple with, let alone value, difference – racial, gendered, sexed, aged, or in bodily or cognitive abilities. Metahumanism is not unique in its disdain for the technicist underpinnings of singularity science and its claims (covert or open) that technological transformation will independently determine the post-human, anti-human, or meta-human future. Nor is it alone in envisaging future forms of future life that are less gross and coarse, less mechanically derived,[3] than those of hard singularity. For many these forms are to be *preferred*.

Katherine Hayles' more or less cognitivist and literary engagements with singularity and the post-human also begin by problematizing transhumanism as a route to thinking about human futures (Hayles, 1999, 2011). Her assessment of strong transhumanism as reductive and determinist in relation to its understanding of the technological, and too narrow and ideologically fraught (bound up with individualism and neoliberal philosophy) to constitute a basis for fruitful discussion, is a common humanities response. Hayles, however, also acknowledges the pull of transhumanism, finding the basic assumption 'that technology is involved in a spiralling dynamic of co-evolution with human development … [a] technogenesis' compelling. The questions the transhumanist community poses about technogenesis in the current era and in relation to human futures are worth worrying about, she declares, not least because they are unresolved. Even so, for Hayles, transhumanism is a catalyst or irritant stimulating more 'considered' and 'responsible' views. These may be developed by way of 'deep, rich and challenging contextualizations' of singularity that 're-introduce the complexities it strips away'. In other words, by exploring them through SF and literary theory.

For Hayles, SF constitutes a better resource to think seriously about advanced technologies and the changes in human lives and culture they produce than anything an exclusive focus on technologies themselves can provide (this is her reading of the singularity community's approach). I would suggest that singularity science/singularity discourse is less purely technological than Hayles' account implies. Partly by virtue of that it is also more important for critical thinking than is often acknowledged. The gung-ho, non-consistent, hubristic, fantastical, heterodox admixture of ideas that circulate as singularity discourse give it a *more* than catalytic function – at least if the latter means it can stimulate (elsewhere) but not produce (for itself) a singularity imaginary of some force. Indeed, contra Hayles' declared intention of transplanting these debates into the 'richer' grounds of literary theory, her own explorations often evidence an ongoing engagement with singularity discourse formations; this is one of their virtues. Her account of Nancy Kress's *Beggars in Spain* (1993), a work that deals in augmented humanity arising through sleeplessness and the time riches it produces, exposes brilliantly how *Beggars* is at once in dialogue with, and in revolt from, forms of the Ayn Rand-influenced libertarianism that pervade transhumanist discourse. My point is that if certain distance from, and disdain for, singularity discourse evident in critical theoretical or literary analytic writing is understandable, given singularity's penchant for declaring the end of any kind of human capacity to make history at the hands of technology, this disdain is also somewhat disingenuous. It is from computing and biotech, and in popular and general-specialist readings of these fields; from a tangle of claims, demands, assumptions, technologies, predications *about technology*, that an energy fuelling claims of post-humanity is found. Speculative and critical theory respond to this energy and to the urgency of the claims it makes, even if in dissenting ways.

SF itself may claim a kind of expertise in singularity. Suggesting this might invoke genre trouble since SF, peculiarly self-aware as a genre, obsessed with making or breaking its own boundaries, has entertained fierce debates about the degree to which it relies on resources from beyond itself. An argument long influential was that adherence to credible scientific logics provided SF with epistemological gravity, a form of authority and grounding that anchored otherwise only fantastical or incoherent worlds (Suvin, 1979). Those challenging this argument (notably Bould and Miéville, 2009) convincingly argued that magical thinking is at least as important to the constitution of SF worlds as science 'itself', something evident given that recourse made to the latter in fiction is always to an *imaginary* science.[4] Once this is accepted, then the conventional divide between fantasy and SF (discussed further later) is breached, as Bould and Miéville (2009) pointed out, and a standard and entirely inadequate model for explaining how SF relates to technological futures by 'inventing' them also falters (Bassett et al., 2013). These arguments have a particular piquancy in relation to singularity discourse, because it is itself, despite its scientism, also and self-admittedly, a discourse partly produced through and *as* (science) fiction.

However, if it is not tenable to hold that SF has a privileged relation to singularity science because it has a deep knowledge of this science, nor because the latter provides an epistemological gravity to anchor the writing, then I nonetheless assert that there is a kind of privileged or over-determined relation that pertains between SF and those questions with which singularity discourse concerns itself.

To build this case what is first necessary is an acknowledgement that SF tends to have an over-determined relationship with transformed materials, environments, or material conditions, although these materials do not have to be mechanical, artificial, or computational. To accept this is simply to notice that SF is a literature of difference and change. Technology co-constitutes our world and organizes and conditions how we lay hold of it – as may magic. This claim enrols an essential genre distinction that contributes to what makes SF (science, speculative, fiction, fantasy) distinctive (even if we accept that genre distinctions are not absolute). It isn't accidental that SF deals in and with singularity questions; this is consonant with what makes it what it is.

Second, it is to be noted that public discourse around singularity is often as much about the magical and the irrational as it is about hard science or technology (conventionally defined). Here I want to briefly invoke research on SF and influence using digital humanities tools, undertaken with Georgina Voss and Ed Steinmueller (2013). This involved following singularity memes on the internet, starting from their fictional 'roots' (e.g. websites of SF authors), through popular science, academic and private scientific communities, and sites promoting alternative belief systems of all kinds. The flow reached singularity organizations, linked the aroma-therapeutic to the extropian, hooked up *Wired* to aliens, and reached through singularity organization sites far into SF communities. What the trails indicate, amongst other things, is that SF is an actant in singularity discourse networks; its fictional claims condition how singularity issues are taken/taken up, contributing to the strength of the network as operational, and the qualities of its operation.[5]

The third leg needed to make the case for SF's privileged relationship to singularity concerns care. A peculiarity of singularity discourse is that key players and insiders declare the stakes to be the highest possible (human existence is in the balance). More, their prognostications are reported in respectable places (the scientific press, for instance IEEE publications) and in mainstream media. Still, however, they are mostly not believed; or, rather, the gravity of their claims is not fully acknowledged or taken on board. Publics in general are not joining the game, committing to the values of the singularity field, which is perhaps why the alarming pronouncements of those who *are* expert players in the relevant fields (Weizenbaum once, now people like Bill Joy, Bill Gates, Stephen Hawking, and others in AI-related fields) have little purchase, their messages regarded as alarmist or sensationalized or overly speculative. Whether these calls are right or wrong is not the issue here. What matters is that *since we don't really believe them, we don't really care* – and vice versa.

SF doesn't have this problem. It doesn't have to be 'taken seriously' to be taken seriously, or for the forms of life, of being, the possible collective futures that it produces, to be cared *about*. What undermines serious discussion of singularity, what produces the paradoxically hyperbolic and bathetic discourse of carrots and upgrades, futures and skin care, serious science journals and sensationalism, brutal dismissal and over-eager hype, loose promises and big money – which is the matter of unbelief – *doesn't matter*, isn't relevant to, SF's explorations of singularity. SF, then, produces grounds in which it is possible to care about a possible future and the material forms that life takes without having to believe in the science that supports the journey towards its implementation, or its feasibility. Routing around the impossibility of caring about what we do not believe in, which produces an ambivalent public response to singularity, SF can provide a *place* to *care* and *ways* to care about singularity futures and about the post or meta or future human or non-human variants it might support. As part of that it can provide ways to explore new forms of being; whether these are then accounted for in critique by way of interpellation, via an account of cognitive estrangement of the *novum*, the attachments of narrative, or the satisfactions of the game, is of less import – fictionality rather than form is key here. Distinctions between SF and 'scientific' accounts of singularity, then, centrally concern how *claims* to ontological truth or fictional veracity or intensity are articulated.

Recent work by both Donna Haraway and Bruno Latour explores care and technological futures in relation to limits to growth and matters of the Anthropocene (Latour, 2011, 2018, Haraway, 2016). Haraway has long drawn on SF's capacity to reconfigure possible forms of future possibility in her critical science writing; her 1980s cyborg was a mythical being, and the late work, treading less lightly, deploys theory-fiction in more straightforward narrative ways – albeit problematically (see Haraway, 2016). The links I want to develop between (science/speculative) fictional figuration, technology and its limits, and care build on both to some extent but also take a different form. Care can be too easily invoked as a term implying it is its own solution – only *care*. But the question is how? In work around the unrepresentable (far futures, far catastrophe, far-flung humans, far-out humans), care has to be invoked in a more demanding way, as what must be undertaken despite what cannot be felt, or experienced, or fully *comprehended*. If care is a responsibility that may be difficult to take on, partly because it cannot purely be a question of 'knowing' – an epistemic question – then how might SF, beyond such issues and debates as those that cohere around the epistemological gravity issue, enable forms of care?

A response might be found by turning to explore the 'greater realism' that Le Guin identified in SF work and said would be helpful for future thinking – including, presumably, for thinking future forms of post or anti-human being – and was therefore sorely needed. However, responsibility in relation to SF is certainly not a matter of pedagogy (nor does it confine SF to what Benjamin termed operational forms of writing). However, Le Guin's comments function to expose a tension in SF, or perhaps an oscillation; SF may be understood through its engagement with the dynamics of utopia – which entails an escape[6] from the boundaries of the given, and a travelling on, a roving, but it may also be explored for its staying power, its capacity to take seriously, stay with, pay fierce attention to, precisely to *take care* of the places it finds itself in, or the places and peoples it writes.[7] To say that SF cares is not to discern a specific orientation, utopian or dystopian, left or right. It is, rather, to suggest that care is something SF can do because of what it *is*; something which might be exploited, or pushed, or engaged with in various ways.

Working through this, I now consider how a series of SF works have taken care with singularity and AI issues and how, in taking care, they have taken sides. These are not technological dystopias, or are not read as such here. There is a long tradition of writing against machine intelligence, from Frankenstein on, and also in film, from the *False* Maria to *Terminator*, perhaps, that could be invoked. My intention here is to exploit SF's capacity to attend, to stay with, to explore more ambiguous formations. These enable an exploration of how an anti-computational impulse is tested in the flesh, how language itself can be anti-computational, as well as, in an odd sense, always already post-human, and how questions arising around singularity scale up. The works invoked here are striking for the ways in which they treat and challenge that urgent/unreal matter at the heart of singularity, one that arises and arises again: human existence versus the rise of machine intelligence. This is a matter we are familiar with these days, although we hear it more as a routine exhortation in relation to devices than in relation to ourselves – upgrade or replace?

The difference that makes a difference: bodies of glass in the 1990s

In the early 1990s cyberspace generated and circulated a new generation of upload fantasies. Hans Moravec notoriously proposed a robot bush as a viable new form of body, welcomed the AI singularity, celebrated the end of the human, and looked forwards to the rise of his new overlords (Moravec, 1988). Singularity featured heavily in William Gibson's early fictional worlds, and also figured within a then extant popular imaginary which valorized or feared the coming virtualization of life, but either way was fascinated by it. Of course, cyberspace was not 'the internet', and the terms were never simply interchangeable; neither the avant-garde imaginaries of publications such as *MUTE*, lists such as Rhizome, nor the milieu of the Institute of Contemporary Art (ICA) nor the popular imaginary and/in its commoditized forms (advertising for the computer industry would be a good example), nor even digital cultural studies, were ever as thoroughly cyber-spatial as has since been suggested. Certainly not if that meant they were gulled by its call to abandon Real Life for the virtual space behind the screen. Nobody jumped in. However, there was the internet itself, and the airy transactions it allowed, fuelling speculation about new forms of being and enabling new ways of exploring it, even if in very partial ways. Further, there was cyberspace as William Gibson wrote it in *Neuromancer* (1984) and elsewhere. This did include the admittedly ambiguous but nonetheless potent hymn to the virtual network, the new other place whose tangled complexity gave it a peculiar substantiality even in its refusal of substance. *This* was the consensual hallucination, the unthinkable complexity, the light space ranged in the mind, and it did provide the beginnings of a way to think about how to live, and what new life might be. Cyber and its new or refreshed suffixes – space, time, bodies, punk – certainly came again to be emblematic of a particular moment.

Gibson has since declared his indifference to issues of the singularity, that possible or fantastic future event, believed in, derided, dispatched, revenant, hoped for as salvation, or feared as the end of the human. Deliverance or death, or that event we may simply find ourselves on the other side of one day and wonder what the fuss was all about. Nor did he ever choose between the flesh and the virtual life. However, a cyber preference was how the cultural imaginary read his 1990s works at the time – and with reason. It was 'meat' that the despairing hero of *Neuromancer* fell back into when he lost access to the network; 'simstim', which involved the simulation of the flesh was rather despised; and there was the bodiless, the *specifically* bodiless, exaltation of cyberspace, those lines of light ranged in the nonspace of the mind (Gibson, 1984). This sensibility was consolidated in more popular cultural arenas, for instance via the later *Matrix* films. It chose the virtual, the artificial, the mind over the body, which is viewed as, in the end, possible to step over, or set aside.

Marge Piercy's early 1990s novel *Body of Glass/He, She and It* (1993) was avowedly written[8] as an *interested* response to cyberspace and the prospect of AI and new forms of life. It set out to rethink in feminist terms what Piercy viewed as the masculine attachments of cyberspace, some of the absolute divisions (on and offline) it appeared to cleave to, and some of its presumptions about technology and everyday life. Responding to cyberspace, Piercy was developing a long-standing feminist engagement with technological societies, artificial life, and gendered possible futures. Her *Woman on the Edge of Time*, a paean for a woman of colour who makes a choice about the shape of the future, despite the powerlessness of her position in the time she finds herself in, is relevant here.[9]

In *Body of Glass* attention is paid to embodied intelligence, human and machinic, to how a person (artificial or not) comes to be, and to the texture of a precarious but possible domestic life. There is less of the alluringly dystopian darkness of *Neuromancer*, more environmental degradation, and less of the bittersweet aesthetic appreciation of the 'television sky' that is part of cyberspace imaginaries. The network is there, but as now as a birthright for those with rights, become infrastructural, with all the unevenness of distribution and access that the term implies. The prospects for intelligent machines and artificially augmented humans emerge through an overlapping series of connected narratives through which themes of perfection, mastery, slavery, and freedom, are developed. The protagonists include a cyborg named Yod, the tenth in his series, a highly enhanced human woman, a partly sentient house invested in the species-purity of humans (and apparently troubled by the cyborg's 'improper' relationship with one of its inhabitants). There is also the golem Joseph, whose rise into life and fall back into clay in an embattled Jewish community is recounted as a tale within a tale, distant from the events of the main narrative, but recapitulating its central dilemma:

how may a life be lived, how can a being *be*, if that life is owned by another, its creator and father; if, by virtue of its constitution, its *purpose* is not its own?

Body of Glass prefers some forms of life to others, which is to say that some lives appear more viable, in multiple ways but particularly in relation to socio-economic arrangements (sovereignty, ownership). Yod in the end sees viability in binary terms – augmented humanity versus artificial life – but the divisions the work articulates are more complex. First, then, there is the fact that Yod's uncertain status confounds any simple binary ontology: organic or artificial, human or machine. This alive but not-human being, who is neither he, nor she, nor feels himself to be an it, is self-consciously aware of having an uncertain claim to aliveness. Sex and gender intervenes, as it has done from Turing on to underscore the complexity of such questions as: 'Is it alive?' 'Is life enough to be a person?' 'Is intelligence enough?' Yod passes the Turing test – now made flesh, made sex, made love. But the issue is not centrally about simulation, a matter of passing, which is always in a sense a matter of trickery, nor even about ontology. Yod emerges and acts in the universe of the novel as a being both more than human and less. His impervious skin lets him move freely in a polluted environment and his strength and 'native' capacity to operate in virtual worlds make him a super-human fighter. He is faster, stronger, quicker at calculating. Simultaneously though, his artificial body categorizes him as *less* than a fully human being in relation to the common body of the community which judges what kinds of bodies are legitimate within it and are therefore given rights within it. Yod's programming binds him to his father/master, who may command him to destroy himself and to destroy others. This makes him, in his own radiation-resistant, superhuman, but also introspective eyes, either a controllable machine or a living commodity; effectively a slave. He may choose to protect the community in which he lives, but it may simply demand that protection as a consequence of the fact that he is owned. To pay him wages would be to undermine the community's (sense of) its right to make that demand, since it would recognize him as a living being and therefore also as a citizen – and vice versa. But even if it wished to accord recognition, how *can* an entity unable to make autonomous decisions, by reason of ownership, being legally but also materially defined as the intellectual and physical property of another, be fully part of common decision making amongst a community of other beings, if that community claims it is free, based on each within it having a voice of their own? Yod, then, is subject to the law of a father, and therefore has no legitimate social being. Questions of freedom, citizenship, rights, which might seem to be supplemental to, or come after, singularity itself, are revealed to be central; being part of singularity's making, its materials, its conditions of possibility. *Body of Glass* exposes ways in which considerations of the viability (or preferability) of new forms of intelligent life that arise cannot rest solely on ontology (human brains, robot circuits, flesh or rubber), nor on the form or substance of a 'life' (mud, code, human flesh, for instance), although these are conditioners, but also on that nexus of relations that constitutes a material political economy within which beings come to be.

Singularity's brusque compression of current life to that of the 'human', to be either upgraded or replaced by the entirely non-human or non-organic, is countered here by the complex entanglements of the tale, not least its exploration of (gendered) familial as well as social structures and their tensioned relationships. Moreover, the narrative takes a different route to that classically taken by fictional robots who feel the constraints of their chains – that of revolt, the rise of the machines. The lines are not to be so clearly drawn. Yod is well over half in love with (at least one member of) the human race, and is perhaps because of that also half embracing death, that which was thought to be a human call, that which singularity wishes at all costs to delay/evade/overcome. Yod's self-adumbration of himself, in his own suicide note, is 'I who may not be alive at all ...'. In it he explicitly declares for augmented humanity over living robots. In this fable, artificial life is made untenable, not because of its substance (machine not human flesh) but by the forms of control that may be exercised over it through its commodity status. This is a bare life at best, perhaps a half-life. It is, however, a life, and this is symbolized by its capacity to dissent. Unlike the golem of the ghetto in the tale within the tale who is laid down and unmade by his creators when his role is done, Yod *does* choose. He takes responsibility for the unmaking of his species/series, binding it into his own violent death. He declares *against* his own form of artificial life and *for* augmented humanity.

Body of Glass mirrors back the binary options common in singularity discourse – augmented humans or artificial life – and in doing so refracts and diffuses them. It points to ways that in practice even a fully artificial life might also, through training, socialization, and education, have to become at least partly (post-)human, in order to become at all. Insisting that 'lives' are made through their conditioning and their struggles and are not simply either a matter of nativity or decantation, Piercy's work disrupts the peculiarly standardized invocations of 'the human' in singularity discourse (which are often reflected in more mainstream arenas). Consider that what is to be improved upon or perfected is adumbrated as baseline human (difference thus being laid aside), and also note that what it might mean to invoke perfection as a goal is rarely questioned, neither for what it might exclude nor for what it might make. Disability studies scholars, amongst others, have extensively explored the implications of perfected humanity as a demand in strong AI and have critiqued it in these terms (Brent, 2012). Within singularity discourse itself divisions are evident between those who argue that transhumanism is based on a desire for the final perfection of the human, a 'more complete victory over human limitations' either through AI or augmentation, as Blackford has put it (Blackford, 2013: 424), and those who want to think less in terms of outcomes and more in terms of processes[10] whose endings may not be called in advance.

Piercy, asking what might be good enough for form of life, or what form of life is good enough to be counted as life, is clearly interested in the process and evolution of change, and *Body of Glass* leans towards forms of life that may develop in process, towards subjects able to go with the grain of indeterminacy; the pursuit of perfection in the production of artificial life, notably the robot series of which Yod is a part, is configured as a dangerous and hubristic endeavour. Against this are the successful cyborg adaptations of others figuring in the tale. This is very different from prioritizing a purely human mode of being, but, in the end, I'd argue that Piercy, like her character, prefers the flesh and is, in this sense, anti-computational.

In *Body of Glass* what it means to fall short of 'being fully human', or even having full life, is exposed as deriving from something more complex than 'how advanced' a pass in the simulation game can be achieved, or even how much (how perfect an) AI can be cultivated. Being *fractures* and responses to questions about life's limits, and of its possibilities to expand or grow, take new forms and shapes, or be made in new materials, become partly a matter of perspectives – ideological, normative, contestable. Piercy avoids the cyberspace sublime, but mostly also the explicitly weird promise or threat of monsters. She deals, rather, in quasi counter-factual histories of the future, exploiting a form of realism that resonates with already existing human lives, scales, organizations; this is the thing that might be going to have happened but did not (yet). The twist in the tail (the 'greater realism'?) comes when it is realized that the 'factual' *countered* in *Body of Glass* is the less-than-factual cyberspace and virtual life imaginary that dominated public internet discourse of the 1990s.

What language may speak that code may not

Let's talk about monsters. The monstrous is anti-computational. It refuses obvious rationality. Its affordances solicit not attempts to compute but reactions that may include wonder or terror. The monstrous does not offer solutions, but it might raise, make graspable, or embody real impossibilities. Moreover, even if the monstrous is anti-computational, its materials may include everything artificial. Drawing on tensions between flesh and machine, many such monsters have crawled into the horizon of our own experiences; the False Maria Machine mensch of *Metropolis* in all her robotic femininity and artificial guile stands as an ur-figure for this, and also indicates its conventional gendering – something later cyber-feminists turned around, claiming a ground zero, an uncountable, a not-one.

When the meta-human manifesto declares monsters are promising it aligns itself with visions of a form of post-human life predicated on the dissolution or confusion of human boundaries through their admixing with the computational which also refuses the closures, or the reductions, that this might imply. In the *Manifesto* (and other similar) theorizations, the monstrous stands as a figure for the coming forms of new life that its authors assume are arising by virtue of computational developments currently being systematically exploited. It's not only flesh monsters, then, but monstrous machines, and the *promise* of monster machines, orphans in two directions, as Haraway's manifesto (1991) almost put it, that *matter*.

The fantastic, as a literary form, partakes of this promise, indeed prefigures its theoretical uptake, and certainly produces monsters – from Lovecraft's objectionable weirdness on. Fantasy finds room at least for the weird, the fantastical, and/or the monstrous, which may be explored through creatures that are synthetic as well as organic, machinic as well as fleshy. Sparking off fantasy, and weird, but also going beyond its narrower generic confines (themselves disputed) is the New Weird. It too tends to refuse the abstractions and the closures characteristic of computationally based approaches to artificial life, at least where this stands alone or is valorized for its purity. The underpinnings of Weird as a genre (both considered in historical terms, where Lovelock is an inescapable figure, and in relation to the formal characteristics of Weird writing which inhere in its style and its subject matter) are the unreasonable and the non-comprehensible. Here be monsters, not rational creatures. Here be beings in *uncountable/unaccountable* dimensions, whose reality nothing can capture; the uncanny and the singularly strange.

The Weird is already a revolt in the flesh and it is hostile to the conventional determinations of the technological/artificial. The New Weird, redefining this genre, constitutes another more or less conscious response to cyberspace as a dominant technological imaginary arising around the net, celebrating the disembodiment of the human soul, and foreshadowing the rise of new intelligence streams in/or as purely informational architectures. However, in contrast to Piercy's careful exploration of new forms of life, which relies on an internal consistency, and on the feasibility of new bodies/new life forms' actions, the new Weird operates in unreal modalities of body, flesh, brain, engagement, interaction, recognition. In the case I want to explore it also operates through language, in a story in which unknowable 'monsters' come to speak in new ways and in doing so celebrate the impossible modalities of human language.

The New Weird resonated with gathering disenchantment with the radical potential of the old net, and the model of disembodiment it offered, found on the left. Frederic Jameson once differentiated fantasy from SF by dividing bodies from machines and declaring the former to be technically reactionary; interesting exceptions, cases where the scent of history could be nosed out even in the fantastical, only confirmed this rule, he argued (Jameson, 2005: 60). Contra to this, by the 1990s cyberpunk was often reactionary *because* technicist, capable of storing the neoliberal visions of technological futures and transformations that drive the market, but not able to configure other more radical forms of future desire, nor to deal with how these might be organized, developed, and exploited. The New Weird sets the monstrous against the application of computational exactitude and its extensions into probabilistic prediction based on big data as a marketing or behavioural tool. In this way it can be read as a critical response to the enclosing grid and abstractions of the virtual turn.

On the other hand, it wasn't then (and certainly isn't now) only those with a critical view who were dissatisfied with the inadequacies of an informational vision divorced from embodiment and its impurities. Silicon Valley's corporations have long sought to capture the previously unquantifiable and elusive, the impure as well as the determinate, the material as well as the abstract forms of social life, of life itself. A shift in focus towards the inexactitude of the flesh enables critique but also responds to, and relates to, a more general shift in computational developments and industry goals, from digital to bio-digital as the centre of attention,[11] from the capture of the easily quantifiable in zones already virtual, to the hunt for ways to undertake the deep capture operations necessary to enable the extension of the commodification of the social world that pervasive mediation, big data, and machine learning undertake. The New Weird's fascination with the impurity of hybrid bodies, with the incomputable, in this way joins with, is part of, a more general turn.

But this is still too tidy. China Miéville, a notable Weird practitioner and also a left theorist, interviewed in 2011, argued that the New Weird spoke to the (then) contemporary because it could speak to that obscurity in relations, that mystification, that is at the heart of the market: its fetishism. In particular, it responds to that mystification that is promulgated through the various promises of the computational, made in discourse and expressed in its operations, so that the real issue is computational capitalism (McDonald, 2011). Miéville's claim, therefore, which again relates directly to Le Guin's sense of SF's capacity for greater realism, is that the fantastic mode can engage with lived reality of modernity in ways that realist fiction currently can't. It is a 'default cultural vernacular' (Miéville, 2002) because it resonates with what is also a fantasy, the fantasy of real life under capital, the fantasy that is at the heart of our material world and its emerging technologies. The fantastic, as a literary form, and in particular the New Weird, can (potentially) get at what is itself a fantastical colonization, and deal with it better than conventional forms of literary realism, because it can better get at 'things' beyond natural realism, or empirical realism. Unlike the postdigital, it continues to give technological inauguration and change its full (over-full, impossible, *monstrous*) attention. Unlike object-orientated forms of new materialism, which have flirted with the Weird (see e.g. Harman, 2010), attracted by its attention to the presence of things, however, Miéville accepts the constitutive and performative force of the entanglement between the symbolic and the real; that indeed is *how* he builds his monsters.

Jameson explored Miéville's work in relation to *Perdido Street Station* (2000), but the focus here is on *Embassytown* (2011), which turns on a war in language, in which language itself, in its materiality and as it symbolizes, is a particular kind of weapon. *Embassytown* celebrates the incomputable excess that defines (human) poetic language, and it works through ways in which forms of language and forms of embodied life – whether these are human, alien, post-human, artificial – are bound up with each other.

There are three languages and three kinds of being in *Embassytown*, a human bubble on an alien world. The humans have a polysemic language that refers beyond what it names. The Ariekei have a language that names what is directly and cannot say what is not (perhaps therefore they are enslaved in language, or perhaps they are free of alienation, this is never entirely clear). Then, appearing less often, there is the speech of an artificial being, an automaton whose output sounds human but is also always encoded. Neither of these last two languages may truly inaugurate. One has soul[12] in it, but no capacity to make new worlds, whilst the other enables communication in the formal sense, but does not – perhaps – host life. In contrast, seductive enough to start a world war is human language, exemplified by metaphor, defined in *Embassytown* as the capacity to make new, to say what is not, or *what was not before it was said*. The philosopher Paul Ricoeur[13] has considered metaphor's links to what is (able) to be, arguing that metaphor forces together two terms, and does so around the verb 'to be' (this *is* this – even when it *is* not). It might thus be said to have a kind of 'ontological vehemence' (Ricoeur, 2003, Martinengo, 2010). Metaphor is a lie that makes what is made through the impossible couplings it produces real; the City is a Heart, insists the protagonist of *Embassytown* Avice Benner Cho (ABC), and the 'heartish city' is thus made. The Ariekei want to lie. At Bakhtinian-style festivals dedicated to this, the lie which they may not speak is approached via the ecstatic, the glossolaliac, the trance, the glitch, the stutter, the break; a kind of linguistic practice that is, if not anti-computational, then resistant to its operations, alien to its closures, an efflorescence, a making, a cheating, a simulating that seeks to cheat its way into the real.

Avice, who speaks of metaphor, as a child became a living simile for her alien hosts. She is 'the girl who ...'. She has enacted what they may not otherwise grasp, since in their language they may not say what is not. But the enactment undermines itself. *Being* a simile, Avice is traducing simile's claims to only *being like*. She *is* the girl who ... The distinction between simile as simulation (like) and metaphor as being (is) is both absolute and thus continuously undermined.

Imitation (being like) and simulation reintroduces the third kind of being, and the third kind of language found in *Embassytown*. Alongside the alien hosts and the squabbling humans are machines, including the automaton Ershul, who 'wasn't human, but was almost'. The personhood of Ershul is initially defended by Avice through an appeal to the category of friendship, with its non-fungible demand – 'spending time with most automa is like accompanying someone brutally cognitively damaged, but Ershul was a friend' (Miéville, 2011: 43). As the Ariekei fall into the delirium of language Ershul becomes increasingly silent. She loses her place in the story, becomes 'a character without a plot',[14] an enigma, an absence, some(thing) already gone. Her liveness itself appears compromised, her two-dimensionality comes to matter – 'her avatar face froze, flickered and came back on' (Miéville, 2011: 115). She seems unable to speak of certain things. Ershul becomes a blank or reflecting screen, a television surface. Avice, the alphabetical human, is left wondering if what appeared to be friendship was in the end simply an optimized interface, friendship's communicative affordances affording maximum efficiency for the data-collecting habits of a machine (Miéville, 2011: 188),[15] but amounting in the end only to the *simulation* of a particularized form of interest; logical enough if the simulator was never more than the simulacrum of a person. Exploring code and language, N. Katherine Hayles invokes Kittler, who argues that the 'dilemma between code and language' produces an irresolvable conflict. This means, he says, that 'the program will suddenly run properly when the programmer's head is empty of words' (Kittler, 2008: 46). In *Embassytown* this is inverted. When the humans make friends with the alien hosts, when the vitality of human words becomes an infection, it is the program named Ershul that stops running. When language came, the AI did not speak.

In *Embassytown* symbolic language with its informing vitality and with the capacity to inaugurate that arises through polysemy is monstrous and excessive. It lets in a new form of life; the city is to be remade, and if the shared language it will speak was originally human the materials of the city are thoroughly alien. The war in language in *Embassytown* thus produces something that might aptly be explored as a new collective social body, one which is not entirely human. This might be considered as a form of meta-humanism, perhaps, but it gestures towards something different from the version elaborated in Ferrando's account, not least because, whilst post-human anatomically – since in *Embassytown* symbolic language relies on life itself but not exclusively on any innate human-shaped capacity, cognitive or otherwise, to speak – it remains fleshy and attached to organic being. It comes *into* being by excluding the only artificial being present. So it is through and in language as it is fleshed and voiced that questions of simulation, simulated life, and (language and) recognition are argued out here. And it is the language of metaphor that is furthest from the restrictions of code that speaks against the computational, offering in the place of code which reduces and places and accurately describes, a language that works by lying, exceeding, by falling short, and over-reaching. Language here becomes revolutionary.

The New Weird with its monsters, and its monstrousness, and its delight in impossible excess, certainly replies to the faded and tattered imaginary of virtuality, and in doing so can question a particular form of modernist rationality. I want to stress finally, here, that there is weird, and weird, and weird. Miéville's Weird is at odds with the kind of weirdness loved by the OOO new materialists, particularly as this was hitched to the new aesthetic with its invitation to luxuriate in the strange objects around us, to listen for the pixels to hum, and to forms of media materialism which repudiate meaning. By contrast, Miéville's is a weird in which meaning and mediation are key, and in which humans are always to be implicated in their environments, part of the promise and part of the threat. This form of Weird contains a contestation for the present. If Piercy's 1990s counter-factual realism undermines or challenges the allure of the computational excess, the artificial as fetishized in cyberspace, Miéville's Weird, as it took shape in the early 2000s, celebrated the dense infolding of the flesh and, continuous with that ,celebrates language and its generative capacity. *Embassytown* at least prefers the excess of signification to the determination of code, and in this way prefers particular forms of (new) life over others that are entirely mechanical.

Quantum filth, merged minds, residual humans ...

Finally I turn to the *Fractal* cycle of H.R. Rajaniemi, a Finnish mathematician who writes in English. The *Quantum Thief*, the *Fractal Prince*, and the *Causal Angel* (published between 2010 and 2014) explore the stakes of singularity and the transhuman across the grounds of a baroque universe over-stuffed with multiple forms of human and post-human life and after-life. The gravity of terrestrial worlds is mostly set aside in favour of exploring a post-singularity Space-opera universe characterized by the generalized extension of the debt relation to include the literal accumulation of life itself in a universe of multiple intelligence streams.

In the *Fractal* novels contestation is not over language but, rather, follows the logics of serious gaming; Huizinga peeks in, rather than Ricoeur. A series of warring parties, each adopting their own form of post-human life, strive against each other, or occasionally cooperate, to survive, thrive, dominate, or overcome. This is a game played in the aftermath of singularity. The protagonists include a gaming society, the Zoku, committed to matter and championing a form of entangled existence formally allowing individual consciousness, but also operating through shared and increasingly potent volitional 'nudges' modulating individual desires. Against them stand the endlessly acquisitive Sobornost, giant collective brains directed by

Primes or the Gods in the clouds and supported by myriad hierarchically (genealogically) organized clones and billions of collected and uploaded Gogol souls. This is the meat hate side and it appears to be winning.

Between these major players are the exhausted remnants of once more or less baseline human beings, who still adhere to forms of embodied and individual life, albeit in forms highly technologized and augmented and involving compromises with the quantum. These include the Aun, hanging on to old Earth, now a desert corrupted with wild code where couplings with other intelligences can maintain a form of human life, and the Oort, ice carvers who embrace the Dark Man of the vacuum in spaceship saunas (the Oort are thinly disguised Finns) or joining the uploaded. Finally, there is society of the Oubliette, where life is lived on two watches and the fully living count down their life credit and amass more through long work as a Quiet. In the Oubliette, to be fully oneself, flesh meeting the memory, a social (time) debt must be paid.

But this is more than a fairy story turned Space opera; a political formation and a form of post-humanity are here tightly enmeshed. Singularity, and its economy, the debts it produces when bodies and minds become tradable commodities (or futures), is both enacted on bodies (or negates them entirely) and becomes a system. Debt is at the heart of social relations in the *Fractal* universe, and debt relations organize the story arcs of many protagonists. The whole can be read as an inquiry, a series of thought experiments, into how relations of debt *scale* up, and scale *out*, how they translate into dimensions and societies far beyond those lived by humans in terrestrial societies, where domination and the forms it may take are remapped. Debt here, defined as a key relationship in computational capitalism, and already relating to biopower, here links very tightly to the capture of, and quantification of, being. This is singularity as a commodity relation.

Central here are the Sobornost, for whom the entanglements of the physical world are corrupt and unclean. Their war is against matter, their endless desire is to salvage consciousness. All that was human shall become 'gogol' – and of course it was Gogol who wrote a story about the sale of dead souls, a pyramid/Ponzi scheme dealing in never-to-be-delivered futures. Salvage is commercial. For the Sobornost, the 'great common task' of uploading and collecting intelligence constitutes a form of primitive accumulation, one in which a certain kind of immortality is given in exchange for the end of autonomy, individuality, and free will – which is to say the end of the self. Wikipedia tells us that the original Sobornost, a Russian movement that lauded the 'spiritual community of many jointly living people' prioritizing the whole over any form of individuality, was defined as a moment of change. In the universe of the *Fractal Prince* the rise of the Sobornost constitutes that moment, signalling the establishment of a new form of life, the end of embodied and individual existence as default, and the expansion of a mode of primitive accumulation to produce ownership – or an indebted relation – without end.

Work exploring debt and its relation to capitalism has proliferated. David Graeber's (2014) work explores its history. In work exploring (Foucault's) consideration of the relationship between war and governance as strategies of domination, Lazzarato follows some of the same lines but specifically explores war and debt. Lazzarato understands debt/credit to operate both 'as a dispositif' or form of governance and 'as war, that is to say, as a strategic confrontation ground' (Lazzarato, 2011: 3). His position is that 'telluric forces' of 'deterritorialization', or accumulation by dispossession, are now continuous. In *The Making of the Indebted Man* he argues that as a consequence everybody owes (Lazzarato, 2011: 1). This also means everybody is owned, or rather that the vast majority of the population are indebted to, or owned by, the very few who are the creditors. Moreover, they are owned into the future, which is to say their future is owned or foreclosed, because they are in debt for their time to come (Lazzarato, 2011: 32).

In the *Fractal* universe, credit, debt, indebtedness, and foreclosure operate at huge scales. A new feature, however, is that debt tends to become permanent, singularity terminating that which might have wiped the debt clean (at least for the debtor) by confounding death itself. Through uploading, indebted being becomes a quasi-ontological state (perhaps as well as a legal or heritable one). Subordinate clones are always indebted to their makers; this is an indebtedness without reason, not acquired through inheritance. The cost of uploaded immortality, dividing consciousness and perishable flesh, is the condition of indebtedness *going forwards*.

Lazzarato's argument is that in the end war is not over territory but population, and if war turns back into a matter of governmentality it is still population that remains the key. Chiming with that, what is at issue in the *Fractal* universe and the wars it explores is the *matter* of the population; its consciousness (to be subsumed), its body (to be disposed of), its very life, which is to be accumulated, or brought into the creditor/debtor relation.

The great common task of the Sobornost is to abstract intelligence from its chiasmic grounds in embodied life, as a strategy of domination. The projected/desired fulfilment of this process is, perhaps, the end of war, and the institution of a new form of absolute governance – so absolute as to absorb entirely the minds of planetary populations. The end, then, is not discipline (or love, as it was in *1984*), nor even control, which needs after all to be exercised, but terminal recruitment and collection.

In this universe individual players can only try to tilt a board in which massive power is in the hands of the controlling brains powered by their billions of uploads with the cognitive capacities to burn out stars and build them. They include a thief, a goddess (a splinter from a prime), a ship, an ice warrior, an amnesiac child god, and a living ship.[16] They too are entangled in debt relations. The lead is Jean Le Flambeur, a thief much reduced from an earlier more augmented state, personally in debt to the people who spring him from a prison, running endless and endlessly violent iterations of prisoner's dilemma games, who subverts creditor/debtor relations to get what he certainly has not the credit to get. Le Flambeur's machinations are directed not against the particular life forms he encounters but against the disciplinary and regulatory regimes these forms make it possible to impose. This is singularity as political thrill and political thriller.

Le Flambeur exploits the constantly reversing plane of governmentality and war, moving between the scales of conflict and of governance and *messing* with their relation. He is the cheating player on a game board set out by forces far larger than him, he raises the tactical to the level of the planetary scale. Against the harvesters, finding new moves, or undermining given instructions, he represents the puny, but insistent and tricky, finding strength in matter and its complexities, in its irreducible quiddity, in the unique rather than the cloned. Quantum filth against the Primes and their billions of clones, and their consciousness banks, and the rest. In the *Fractal* future at least, in the end, hope is (post-)human.

The novels evidence a contempt for the cold fusion of absolutes that is represented by the terminal uploading of human minds, undertaken to build cognitive power and establish an empire.[17] Continually preferring the continued entanglement of matter and intelligent matter against untethered minds, they disturb any presumption that a cosmically expanded consciousness, or even a hugely augmented mind, naturally constitutes progress. On the other hand, the *Fractal* universe is often joyous. What is rejected is not change or human evolution or new forms of intelligence. Rather, what is mocked is a mode of accumulation producing forms of life capable only of parodic repetition; the absurdity, the hubris, of endless life and endless self, endlessly reproduced, of attempts to exist everywhere and for all time and in a million places at once; absolutely abundant, absolutely the same, and always unfree.

What remains is an affirmation of the non-computable self, still able to surprise, to engage in the exploit, and an affirmation of the way in which this more or less human self has also drawn closer to and is becoming entangled with other forms of being. Le Flambeur and a Goddess who loves him share a form of laughter, a kind of fatalism that frees both to act, that isn't quite human, but isn't entirely *inhuman* either. In the *Fractal Prince* series there is always a beautifully ordered archive, a prison for minds and bodies whose barbed wire is secured far into the past and future; this is the shape the debt economy takes in a world where the currency is self and mind. If there is hope, it is because this form of currency may carry an excess charge, something not quite calculable in advance, and therefore not quite fully integrated into the post-human, universe-wide, debt economy, *going forwards*. To hope for a door is partly to create the possibility that one might exist in a future that is, by virtue of that hope, not entirely owned.

Conclusion

The works explored here span over three decades. They indicate an entanglement between singularity discourse and internet and new media technologies and (increasingly) between singularity and biotechnological cultures. Each is media-archaeologically significant, partly because each complicates or dissents from key narratives circulating around the time of their production. Together they undercut the temporality of the discourse of real singularity with its varied, but strikingly concrete, predictions of coming change. They do this by exploiting the multi-directionality/dimensionality of SF which is both here now and unreal, not only to question what singularity might possibly deliver, but to question the terms, the political economy, of that possible delivery.

They have in common that they exhibit a preference for a continued engagement with matter, and a desire to remain or retain something human. They value matter and/in its relation to consciousness. More than that, they value matter, and human matter, because of its quiddity and resistance to terminal reconciliation. They were chosen to do this, of course; selecting them, I was perhaps treasure hunting, something I have deplored elsewhere. My excuse is that I did it to *counter* another other form of treasure hunting: that which seeks in SF an endless form of affirmation for the fictional future we are given as real by contemporary technological society, a system that 'liberates' certain fictions 'to rule over the social', as Mark Fisher put it (see Fisher, 2009: xii), but refuses others as pure fantasy; and that does this in relation to technology ruthlessly and relentlessly. Like Le Guin, I think SF can develop a much-needed critical and realist engagement with what is given to us as the reality of the bio-computational future. This is why, as Mark Fisher noted, it may not only simulate, elsewhere, but also come (back) (here) to act now (Fisher, 2009: xiv).

Notes

1 'I have had a long career and a good one ... The name of our beautiful reward is not profit ... it is freedom' (Le Guin, http://blog.worldswithoutend.com/2012/12/gmrc-review-always-coming-home-by-ursula-k-le-guin/).
2 Ferrando argues for a philosophical, cultural, and critical post-humanism, in her case inflected by Heidegger and versions of appearance (Ferrando, 2013: 27).
3 A key thinker here is Braidotti.

4 Suvin (1979) refers to the importance to SF of epistemological gravity arising through the cognition effect and the *Novum* (the creation of worlds, and that which produces estrangement).
5 See Bassett, Steinmueller and Voss (2013). Work commissioned for NESTA explored the relation between SF and 'real innovation'. Our methodology was informed by Franco Moretti and moved between hermeneutic and computational analysis.
6 Oscar Wilde (2001) famously said of utopia, 'when humanity lands there, it looks out, and, seeing a better country, sets sail'.
7 This kind of caring being at its most careful light years away from the dogmatic commitment of overtly pedagogical writing.
8 Piercy invokes Gibson and cyberspace in her acknowledgements in *Body of Glass* (Piercy, 1993: 583).
9 *Women on the Edge of Time* (1976) was striking for its engagement with a feminist utopian society in tune with its environment but under threat, and barely sketched the outlines of its dystopian other – a highly techno-celebrant society, the horrors of which are only glimpsed.
10 See, e.g. Blackford who is in favour of a 'more complete victory over human limitations' versus Martha Nussbaum's sense that a 'more than human life would not be a good life' (Blackford, 2013: 424–425).
11 The rising market for biotechnological consumer products is one indicator of this trend, amongst other things.
12 An impossible, incomprehensible language. When the Ariekei speak, says a linguist in *Embassytown*, 'the words have got soul in them' (Mieville, 2011: 56).
13 A 'Paul Ricoeur' is named in *Embassytown*, one of the philosophers whose linguistic theory might help explain the alien language. If the invocation of a real philosopher constitutes a metafictional appeal to plausibility of the linguistic possibilities explored, then it might be justified to argue that it is linguistics rather than physics that provides a form of epistemological gravity for this novel.
14 As an internet fan noted.
15 'I'd been to her home. Tasteful accoutrements, perhaps for mine and others' benefits: elements of an operating system, designed to make her user-friendly. These ruminations felt disgraceful' (Mieville, 2011: 103).
16 A living ship, and then a missing ship named *Perhonen*, which is the Finnish word for a butterfly.
17 If these works are exploring the contest against a rising empire where the odds are stacked in the empire's favour, perhaps this shouldn't be a surprise. This is, amongst other things, a Finn writing about a Russian-named project of absorption, invasion, and empire.

Bibliography

Bassett, Caroline. 2012. 'Why after Language Came the AI did not speak'. Paper to Weird Council: An International Conference on the Writing of China Miéville, September, Birkbeck, London.
Bassett, Caroline, Ed Steinmueller and George Voss. 2013. *Better Made Up, Science Fiction and Innovation*. Working paper to NESTA.
Blackford, Russell. 2013. 'The Great Transition: Ideas and Anxieties', in Max More and Natasha Vita-More (eds) *The Transhumanist Reader: Classical and Contemporary Essays on the Science, Technology, and Philosophy of the Human Future*. London: John Wiley and Sons, 421–429.
Bould, M. and C. Miéville (eds) (2009) *Red Planets: Marxism and Science Fiction*. London: Pluto Press.
Braidotti, Rosi. 2013. *The Posthuman*. Cambridge: Polity.
Braidotti, Rosi. 2015. 'Posthuman. All too Human? A Cultural Political Cartography'. Lecture at UEL, https://rosibraidotti.com/2018/06/29/posthuman-all-too-human-a-cultural-political-cartography/.
Brent, Walter Cline. 2012. '"You're Not the Same Kind of Human Being": The Evolution of Pity to Horror in Daniel Keyes's Flowers for Algernon', *Home*, 32(4) (no pagination).
Butler, Judith. 2002. 'What Is Critique? An Essay on Foucault's Virtue', in David Ingram (ed.) *The Political: Readings in Continental Philosophy*. London: Basil Blackwell, 212–226.
Čapek, Josef and Karel Čapek. 2010. *R.U.R.: Rossum's Universal Robots*. London: Wildside Press
del Val, Jaime and Stefan Lorenz Sorgner. 2011. 'A Metahumanist Manifesto', in *The Agonist: A Nietzsche Circle Journal*, IV/II, www.nietzschecircle.com/agonist/2011_08/metahuman_manifesto.html.
Ferrando, Francesca. 2013. 'Posthumanism, Transhumanism, Antihumanism, Metahumanism, and New Materialisms: Differences and Relations', *Existenz Journal*, 8(2) (no pagination).
Fisher, Mark. 2009. *Capitalist Realism: Is There No Alternative?* London: Zero Books.
Gibson, William. 1984. *Neuromancer*. London: Gollancz.
Graeber, David. 2014. *Debt: The First 5000 Years*. London: Melville House.
Haraway, Donna. 1991. *Simians, Cyborgs, and Women: The Reinvention of Nature*. New York: Routledge.
Haraway, Donna J. 2016. *Staying with Trouble: Making Kin in the Chthulucene*. Durham, NC: Duke University Press.
Harman, Graham. 2010. *Circus Philosophicus*. London: Zero Books.
Hawking, Stephen, Stuart Russell and Max Tegmark. 2014. 'Transcendence Looks at the Implications of Artificial Intelligence – but Are We Taking AI Seriously Enough?' *The Independent*, 1 May.
Hayles, N. Katherine. 1999. *How We Became Posthuman: Virtual Bodies in Cybernetics, Literature, and Informatics*. Chicago: University of Chicago Press.

Hayles, N. Katherine. 2011. 'H-: Wrestling with Transhumanism', in Gregory R. Hansell, William Grassie et al. (eds) *H+: Transhumanism and Its Critics*. Philadelphia, PA: Metanexus Institute, 215–26.
Jameson, Fredric. 2005. *Archaeologies of the Future*. London: Verso.
Joy, Bill. 2015. 'Why the Future Doesn't Need Us', *Wired*, 22 April, http://archive.wired.com/wired/archive/8.04/joy_pr.html, page 6 of 17.
Kittler, Friedrich. 2008. *'Code' in Software Studies: A Lexicon*, ed. Matt Fuller. London: MIT Press, 40–46.
Kress, Nancy. 1993. *Beggars in Spain*. London: William Morrow and Company.
Kurzweil, Ray. 1999. *The Age of Spiritual Machines*. London: Viking.
Kurzweil, Ray. 2005. *The Singularity is Near: When Humans Transcend Technology*. London: Duckworth.
Latour, Bruno. 2011. 'Love Your Monsters: Why We Must Care for Our Technologies as We Do Our Children', *Breakthrough Journal*, 2: 19–26.
Latour, Bruno. 2018. *Down to Earth: Politics in the New Climatic Regime*. London: Polity.
Lazzarato, Maurizio. 2011. *The Making of the Indebted Man. An Essay on the Neoliberal Condition*. New York: Semiotext(e.), Intervention Series 13.
Lazzarato, Maurizio. 2015. 'Debt as a Continuation of War by Other Means'. Paper at *Debt: Philosophical, Cultural and Interdisciplinary Perspectives on Exchange Relations* (Jane and Aatos Erkko Professor Seminar), Helsinki Collegium of Advanced Studies, 13–15 April. [Original title: The Infinite Debt]
Le Guin, Ursula K. 2014. 'Hard Times'. Acceptance speech for Medal for Distinguished Contribution to American Letters at the 65th Annual National Book Awards ceremony. Video at truthdig: www.truthdig.com/videos/science-fiction-writer-ursula-k-le-guin-movingly-warns-against-the-dangers-of-capitalism-video/.
Martinengo, Alberto. 2010. 'Metaphor and Canon in Paul Ricoeur: From an Aesthetic Point of View', *Proceedings of the European Society for Aesthetics*, 2: 302–311.
McDonald, John. 2011. 'Fantasy, Science Fiction, and Politics. Interview with China Miéville', *International Socialist Review*, 75, https://isreview.org/issue/75/fantasy-science-fiction-and-politics (accessed 16 May 2021).
Miéville, China. 2000. *Perdido Street Station*. New York: Del Ray/Ballantine Books.
Miéville, China. 2002. 'Editorial Introduction', *Historical Materialism*, 10(4): 39–49.
Miéville, China. 2009. 'Cognition as Ideology: A Dialectic of SF Theory', in M. Bould and C. Miéville (eds) *Red Planets: Marxism and Science Fiction*. London: Pluto Press.
Miéville, China. 2011. *Embassytown*. London: Macmillan.
Moravec, Hans. 1988. *Mind Children*. Cambridge, MA: Harvard University Press.
Nussbaum, Martha. 1992. *Love's Knowledge: Essays on Philosophy and Literature*. Oxford: Oxford University Press.

O'Riordan, Kate. 2017. *Unreal Objects: Digital Materialities, Technoscientific Projects and Political Realities*. London: Pluto Press.
Piercy, Marge. 1976. *Woman on the Edge of Time*. London: Women's Press.
Piercy, Marge. 1993. *Body of Glass*. London: Penguin.
Rajaniemi, Hannu. 2010. *The Quantum Thief*. London: Gollancz.
Rajaniemi, Hannu. 2012. *The Fractal Prince*. London: Gollancz.
Rajaniemi, Hannu. 2014. *The Causal Angel*. London: Gollancz.
Ricoeur, Paul. 2003. *The Rule of Metaphor*. Oxford: Blackwell.
Suvin, Darko. 1979. *Metamorphosis of Science Fiction: On the Poetics and History of a Literary Genre*. New Haven: Yale University Press.
Tegmark, Max. 2018. *Life 3.0: Being Human in the Age of Artificial Intelligence*. New York: Knopf.
Wilde, Oscar. 2001. *The Soul of Man Under Socialism and Selected Critical Prose*. London: Penguin.

Filmography

Plug & Pray. 2010. Jens Schanze (director).

Conclusion

Upping the anti: a distant reading of the contemporary moment

In the 2020s we live in an era of automation anxiety and automation fever (Bassett and Roberts, 2020). There are rising concerns around extant computational cultures, near-horizon developments, and longer-term predictions about computational futures. Anti-computing of various kinds is back with a vengeance. A new moment of urgency arises. Once again the contemporary moment is proclaimed as *the* time of make or break, *the* time for decisive action. Once again it is argued that technology will in short order crystallize into a good or bad angel, working for good or ill, as Norbert Wiener put it, back in the 1950s – and his work too has undergone a revival. For a flavour of the moment look at what the books say; *Democracy Hacked* (Moore, 2018), for instance, claims we're in the last chance saloon, 'if we don't change the system now, we may not get another chance', whilst *People versus Tech* (Bartlett, 2018) fears that unless 'we radically alter our course, democracy will join feudalism, supreme monarchies and communism as just another political experiment that quietly disappeared'. Others' fears are more existential, reading the contemporary moment as an AI tipping point.

Reading, or, rather, the matter of what is being read, is germane here. This chapter is informed by sixty-four books that are critical, anxious, hostile, concerned, that are actively writing *against* the computational state we find ourselves in. But this is not a close reading. All the titles were located through Amazon, the majority written within the past five years, but all found in 2020. The contents of some of these inform the arguments here, but the chief interest is in a meta-narrative: how automation anxiety is being framed and sold, how those tropes identified as characteristic of various categories of automation anxiety are expressed, and how they recombine – or divide – in new ways today. Amazon's algorithms thus assisted in a somewhat unofficial form of distant reading (Moretti, 2013). The method – barely one – was simple. I swiped (left and right) across Amazon's quasi-Borgesian 'shelves', letting it tell me what other readers read (or bought) when they started where I started, with the books I started with. I also explored Amazon's 'promoted' reading suggestions, which often threw up titles further from my concerns than the reader recommendations.

Three or four 'seed' books produced the vast majority of the titles. A few emerged on later searches when I used new terms to route around disconnects; cybersurveillance and detox, for instance, didn't quite intersect in expected ways. I make no claims to completeness in undertaking this exercise – nor to having isolated a discrete genre; these are not all anti-computing books, the boundaries of what constitutes a hostile or investigative account are disputable and sometimes blurred, and I certainly have no access to Amazon's algorithms, which occasionally threw up an entirely unexpected suggestion. However, as a probe, this did produce results that were startling, and not least in their quantity. Overall, the search evidenced an explosion of anti-computing taking many forms and evident across a series of genres – I avoided a sharp division made repeatedly, for instance in Paul Mason's *Clear Bright Future*, between 'airport' accounts and 'serious' endeavours. This was a cornucopia of interpretation, apostasy, outrage, sorrow, critique, and anger – along with robust demands that readers suffering from computational cultures buy this book, or that one, to help themselves, or perhaps be *cured*. (Emma Harrison's extant and forthcoming work on connections between mental health and cultures of digital anxiety is germane here.)

The probe thus indicated in outline the rough shape of contemporary anti-computing formations, suggesting key foci, structures of feeling, emotional registers, the orientation and/or bias of currently loudly heard arguments. To consider these formations in relation to the taxonomic categories developed at start of this book is to ask how or if they fit. They may also indicate how a category has morphed, how connection with older or longer-standing formations is made, what of them is really new, and what persists or is revived from earlier moments, or revenant discourses. This is undertaken in this chapter. It leads to a brief reassessment of anti-computing itself, offered by way of a conclusion to the book as a whole. The emphasis is on reconsidering the relationship between computational thinking as a cultural form encapsulating both fatalism about, and a desire for, the further rise of the computational, and the anti-computational impulse, also fatalistic at times, but at others making demands for change. I conclude finally that a certain kind of anti-computational thinking is *necessary* for truly hopeful computational futures to be envisaged, sought – or fought for.

Distant reading swallows Amazon

I am looking for books attacking Amazon on Amazon. The platform obliges with *The Four: The Hidden DNA of Amazon, Apple, Facebook and Google* by Scott Galloway (2017). Amazon is of course largely indifferent to the specifics of content. It doesn't care what the books it sells communicate, it's more interested in what I communicate to its algorithms in my fevered search for hostility to, amongst other things, the platform upon which I conduct my searches. The need to capture that data might produce a desire for a better semantic understanding but there is no need for Amazon's bots to appreciate irony, at least if the latter is measured in human terms, only perhaps to understand that it is there.

Amazon of course began with printed books – and perhaps it is surprising that it still sells them. The book refused to die. You knew this, I'm pointing to it because there's a confluence to be observed; here are objects which could be purely informational, that are, by our own desires, still (sometimes) produced on paper, with cardboard covers, using CMYK (cyan, magenta, yellow, black) inks to absorb or reflect colour rather than relying on RGB (red, green, blue) screen display technologies. Reading paper-based books itself begins to represent a slightly tenacious refusal to upgrade, or to be 'up to date', and Amazon's book pages, suggesting Kindle but just as willing to sell paper, both escalate the omnivorous logics of the digital *and* point to a long-term, non-trivial, intrinsic, attachment to other kinds of material; to a form of resistance. They also point to the impurity of 'the virtual' as an offering in a material world, since even if books finally become fully digital objects there will still be bodies to read them, and this side of authoring developments signalled by OpenAI's GPT2/3 language models at any rate, (post-)human authors to write them. So, although not surprising, there is something *nice* about Amazon recommending (or passing on 'reader' recommendations, however these are manipulated through encoding) books that are hostile to its materials, to itself, to its logics; books that offer ways to break an addiction to the forms of networked sociality and communication that it itself promotes and lives by. Printed books are far less insistent that readers do not close the covers of a discrete work without thinking about the next one 'before you go ...'. In all, then, what better place to stick around in, if we are looking for where anti-computing has got to by the turning of the first two decades of the 21st century; right *now* – print publishing turnaround times – and my own tardy writing – permitting.

What is this now? Or, rather, *when* is it? These are the days of politics dominated by Trump's tweets and Russian interference, Brexit populism, fake news, rising acknowledgement of platform monopoly. We live beyond the knowledge that the Cambridge Analytica scandal gave us, in an era of systematic bucket data capture, concern around screen addiction, trolling, bias, misogyny, addiction, and the rest. Mark Zuckerberg has declared the need for privacy and a desire to step his platform away from the political writ large; a move back to the drawing room, as if that would chase out 'the political' (it's always already personal, a chorus of feminists point out). In the UK at least, there is a sense of political crisis, and a moral panic about the internet as partly 'what did it' – but also a sense that disengagement is not possible, particularly in COVID/Zoom times).

The above of course is a grossly compacted scenario. I invoke it to stress the force with which the sentiment that we live in unusual times (of political crisis, environmental crisis, moral crisis) is being felt. Which is why it is useful – even salutary – to consider how contemporary computational hostility, arising in response to that formation, part of it as well as mapping onto it, challenges and partially reshapes the more general taxonomy I began with, *but also works through its invocation*. Today's anti-computing is a recurrent, partly familiar affair, as well as something new. Many of the tropes that inform it are ready to hand, some in more or less complete form, others it is itself reshaping.

Notes from an annotated meta-bibliography

The general taxonomy of anti-computing developed at the start of this book (Figure 3 and Chapter 1) was generated both by paying attention to real-world formations across time and by working with critical theorizations of the media-technological and of medium-technological cultures. A binary division between ontological and political justifications for anti-computational thinking was identified as informing popular thinking, but also as failing to hold, despite its ideological force. It was also recognized as inadequate as a way of grappling with, or categorizing, these formations in order to critically investigate forms of anti-computing. A more elaborated taxonomy of anti-computing was developed which avoided this division. Eight different forms which anti-computing has typically taken were categorized and potentials for cross-cutting them with other classificatory systems (e.g. affective categories, critical theoretical divisions) were explored.

What is offered now is a partly automated reader's report that responds to these categories. As already noted it works with over sixty books which Amazon thinks somebody with an initial interest in something anti-computational should read *because other readers did*. Following this trail, invoking titles, brief descriptions, and reader comments offered, I explore how these works variously fit into, burst out of, modify, or challenge entirely the original anti-computing categories, which together identified an anti-computational reservoir, a persistent set of tropes, submerged but also ready to re-arise.

---

i. Computer technology as control technology
ii. Computers becoming more lively
iii. Computerization and the hollowing-out of everyday life and social interaction
iv. Computer technologies and the threat to human culture
v. The general accident/catastrophe theory
vi. Horrible humans
vii. Standardization/quantification
viii. Too much information

Figure 3 A general taxonomy of anti-computing.

(i) Computer technology as control technology

Contemporary examples that fit squarely into this category are Jonathan Taplin's *Move Fast and Break Things* (2017), and Jamie Bartlett's *The People versus Tech* (2018). Both make the case for the erosion of democracy by the tech corporations and their products – and both point to the consolidation of power in the new digital economy's heartland. Bartlett's book asks if we have 'unwittingly handed too much away to shadowy powers behind a wall of code, all manipulated by a handful of Silicon Valley utopians, ad men, and venture capitalists?' Taplin considers a shift in the balance of power from old industrial capital to Silicon Valley and the platforms, arguing that 'with this reallocation of money comes a shift in power'. Shoshana Zuboff's *The Age of Surveillance Capitalism* (2019), an academic publication that crossed over to become a best-seller, also identifies 'the threat of unprecedented power free from democratic oversight' of informational capitalism. Franklin Foer's argument in *World without Mind* (2018) follows similar lines as Taplin and Bartlett, with a focus on who 'controls knowledge and information', now in the hands of the 'titanic powers'.

The People versus Tech is introduced by the claim that 'The internet was meant to set us free'. The trope of disillusion or disenchantment recurs powerfully in this category, and with it a sense that freedom was never as present as the hype suggested. Another example is Morozov's 2012 *Net Delusion*. This 'shows why internet freedom is an illusion. Not only that – in many cases the net is actually helping oppressive regimes to stifle dissent, track dissidents and keep people pacified.' Fears about computers controlling humans also arise, although increasingly they fit better in the second category of Liveliness, which considers the *replacement* of the human by the machine (in Haraway's famous 1980s formulation, humans become *less* lively as computers take over more control). Computer power as malevolent or indifferent to human futures remains a strong trope, and one that, despite the rise of new forms of AI and the desire of many AI researchers to avoid the unthinking melding of matters of AI consciousness with matters of AI, remains remarkably stable – and does very often operate on this elision. For non-fiction versions, see later. In fictional SF dealing in singularity and/as apocalypse this blending is something of a staple.

(ii) Computers becoming more lively

Four of the books thrown up by Amazon's algorithms are usefully invoked here, suggesting how long-standing tropes revive but also have evolved. Callum Chase's *Surviving AI* (2nd edition, 2018), lays out the argument that computer autonomy produces human redundancy in existential terms – 'If we get it right it will make humans almost godlike. If we get it wrong … well, extinction is not the worst possible outcome.' Gerd Leonhard's *Technology vs. Humanity* (2016) also fears unthinking AI development: 'The imminent clash between technology and humanity is already rushing towards us. What moral values are you prepared to stand up for – before being human alters its meaning forever?' James Barrat's *Our Final Invention* (2013), meanwhile, explores 'the perils of the heedless pursuit of advanced AI'. These three books exhibit symptomatic concerns and anxiety around singularity – and could also find a place in the first category.

The Glass Cage, by Nicholas Carr (2016), is more narrowly focused. It explores existing automation and its trajectories and considers the likely impact on human cognitive alertness or agency, in an account that stresses the automation of expertise. Amazon says it 'shows how the most important decisions of our lives are now being made by machines and the radical effect this is having on our ability to learn and solve problems'. Human deskilling as the result of augmented computer intelligence is a concern relating not only to human cognitive capacities and their attention, but also to other forms or modes of life. It finds a specific form in contemporary automation anxiety. An obvious reference here, and it is referenced, is Martin Ford's *Rise of the Robots* (2016). This looks at 'the terrifying societal implications of the robots' rise …', arguing that 'any job that is on some level routine is likely to be automated and if we are to see a future of prosperity rather than catastrophe we must act now'; once again, we're in the last chance saloon.

Rather different forms of anti-computing that also inform this category are found, firstly in Rushkoff's *Team Human* (2019), a 'manifesto' which argues that 'society is threatened by a vast antihuman infrastructure that undermines our ability to connect'. The living 'community' of humans is seen by Rushkoff as threatened by increasingly avaricious and active machines. Turning this around,

Meredith Broussard (2018) captures another form of hostility in her look at computational stupidity – or what she terms 'artificial unintelligence'. This returns to an old form of hostility to AI based on critiquing the justifiability of the claims made for it and for the timescale of advance. Like Weizenbaum and Searle (Chapter 6), Broussard's case is that AI does not (perhaps cannot) deliver on the promises it makes.

(iii) Computerization and the hollowing-out of everyday life and social interaction

This category has exploded. Older tropes come back in and inform the framing of new concerns and are, in this process, substantially revived; three forms of currently virulent anti-computing include addiction tropes and (less seen but still evident) other medicalized responses to pervasive connectedness, digital detox as a suggested response, and concern around noise and silence, screen life, and compulsive sociality, which becomes a figure for the lack of solitude or private life per se. None of these tropes is new (they appear in Leavis' accounting with technologico-Benthamism, figure in moral panics around television and family life, reach back to Edmond Berkeley and *Salon*, and more broadly figure in critiques of industrialism itself in all its noise and clamour). However, they are currently being remade, sometimes in radically new as well as in familiar forms, and are expressed as matters of urgency, in registers comprehending anger, sadness – and even despair; reflecting this perhaps are tales of social media and its role in provoking youth suicide.

Of the many books that Amazon throws up for me to pith (since we are hollowing out), those invoked here are chosen because they take a particularly clear line. Psychologist Sherry Turkle's book *Reclaiming Conversation* (2016) is notable for revisionism; Turkle sees it as marking a turn away from the more positive viewpoint she held in her early engagements with digital culture – via *Life on the Screen* and *The Second Self* (2005), although a newly revised edition of the latter is an Amazon suggestion. Turkle now argues that a more critical engagement with the digital is needed. We were young, she says, and so was the internet, but now we need to grow up – there is a 'backlash' (interview in *Vox*, Illing, 2018). The argument of *Conversation* concerns the failure to maintain meaningful relationships across digital media platforms and within a digitally saturated life. A related work is Scott's *Four Dimensional Human* (2018), exploring surveillance and data-captured lives. It is claimed that the book asks 'how do we exist in public with these recoded inner lives, and how do we preserve our old ideas of isolation, disappearance and privacy on a Google-mapped planet?' Michael Harris's *Solitude* (2018) also engages with this – 'In a world of social media and smartphones, true solitude has become increasingly hard to find' – and goes on to delve 'into the latest neuroscience to examine the way innovations like Google Maps and Facebook are eroding our ability to be by ourselves'.

Hooking into addiction tropes, already at hand from gaming, and/or video scares, but less invoked around computer culture/screen culture more broadly until relatively recently, is Mary Aiken (2017), who fears the deleterious effects of everything from 'screens on the developing child to the explosion of teen sexting, and the acceleration of compulsive and addictive online behaviours (gaming, shopping, pornography) …'. Her *The Cyber Effect* 'also examines the escalation in cyberchondria (self-diagnosis online), cyberstalking and organized crime in the Deep Web'. The four horsemen of the internet apocalypse, notoriously predicted by Electronic Frontier spokesman John Perry Barlow in the early 1990s as wreckers of the possibilities for a newly remade society somehow on the edge of old laws (Barlow, 1994), are right here, right now.

Mainlining on addiction is Adam Alter's *Irresistible* (2017), which argues that previously specific kinds of engagement with the computational have generalized: 'Welcome to the age of behavioural addiction – an age in which half of the American population is addicted to at least one behaviour. We obsess over our emails, Instagram likes, and Facebook feeds; we binge on TV episodes and YouTube videos … Millennial kids spend so much time in front of screens that they struggle to interact with real, live humans' – so says the blurb. This book is much invoked for its account of how Silicon Valley parents keep their own children away from the machines they design and market. Will Storr's *Selfie: How the West Became Self Obsessed* (2018), meanwhile, explores and attacks our addiction to ourselves via selfie culture. As Storr puts it, 'our expectation of perfection comes at a cost. Millions are suffering under the torture of this impossible fantasy.'

Finally, there are the detox guides. For instance, Marchant's *Pause* (2018), bringing an 'important message ready to be heard', one whose time is right; 'We check our phones an average of 221 times a day, we have apps that help us sleep and remind us to be mindful whilst we secretly measure our success in "likes".' *Pause* says we should stop doing this stuff. *Time to Log Off* (2017) brings news from a new anti-computing campaign. It is claimed that '[t]his canny little bible will help you log off and wake up to less stress and more time. Enjoy real experiences, real connections and real happiness. Reset your boundaries with carefully crafted exercises, new outlooks and wise words from Tanya Goodin, digital detox specialist and founder of *Time To Log Off*.' Goodin, like Matusow (see Chapter 3), might be said to have a 'gimmick'. Tero Karppi argues that logging off isn't that simple. Her *Disconnect* (2018) reminds us that 'Facebook see disconnection as an existential threat – and have undertaken wide-ranging efforts to eliminate it'. The argument is that users' ability to control their digital lives is, even if they wish for control, gradually dissipating, so that personal anti-computing tactics may not be effective enough these days.

(iv) Computer technologies and the threat to human forms of culture

Some of the older tropes expressing these concerns appear somewhat submerged currently and/or have morphed considerably. An old hostility to gaming versus reading, for instance (e.g. Stallabrass's 1996 attack on interactive multimedia as a shift from the illusion of scene to the illusion of action), has morphed into a concern around excess screen time in general. This produces new links to addiction tropes (often nascent in gaming critiques), particularly in relation to critiques of addiction by design, or the tactics of the software companies to build in the demand for 'more' as part of the attention economy. An example is the invitation to keep binge watching; it is not only in gaming that there is to be no game over (see e.g. Lanier, 2018). There is a more general critique of gamification – which links less to creative activity in discrete zones than to the gamification of everyday life or the reduction of experience to the dashboard. This is implicit at least in Rushkoff's work (see earlier) – and though it didn't arise in my search it is also treated with in critiques of bureaucratization such as that by David Graeber (2015). Striking hostility also arises around gaming, but less around the form than around gaming as a cultural practice mired by discrimination and gender hatred; gamergate – when members of the gaming 'community' attempted to vote down forms of SF whose rise was perceived to be bending the knee to liberal feminism (see Quinn later) or women – is invoked. A different set of anxieties and hostilities arising (with vertical take-off) around cultural production include deep-fake issues. In part, these revise and rework old Photoshop debates arising in relation to news photography and point to a longer-term anxiety about authenticity and 'the real' – in relation to computer art and music, and in relation to fake news and/as public culture (discussed further later).

This category might also encompass hostile responses to (the lauding of) our own remaking as cultural producers. Concerns of this kind are voiced by Andrew Keen, whose 2008 *The Cult of the Amateur* was invoked despite its relative age. This argues that 'much of the content filling up YouTube, Myspace, and blogs is just an endless digital forest of mediocrity which, unconstrained by professional standards or editorial filters, can alter public debate and manipulate public opinion'. Written more or less as a response to the moment of Web 2.0 and its promise not of freedom, which came and went with Web 1.0, but of freedom of production and the claimed democratization of creativity, it appears less relevant to contemporary debates.

(v) The general accident/catastrophe theory

This category gathered together fears about computers spiralling out of control. It is concerned with computer power and complexity and unforeseeable risks. What appears to be emerging in relation to this category is a shift not so much in the magnitude of the perceived threat computation might pose, but in its articulations. Discourse around the Anthropocene, and more generally critiques of untrammelled expansion and its already feared and actually irreversible impacts now more clearly intersect with a particular kind of discussion of computation as part of (disastrous and unchecked and unexamined) growth agendas. This is the background informing writings critical of the degree to which computing is – despite its apparent lightness – energy hungry and environmentally unfriendly, particularly for workers in the global South, as well as for the natural environment in its totality. A recent example of this, slightly early for my Amazon trawl, might be Miller and Maxwell's (2020) work on the mobile phone.

There is also the sharp revival of older tropes framing automatic weaponry; once chiefly expressed via *Terminator*-style fantasies, these are now critiqued and explored in relation to nascent and actual real-world weapon systems, and/in their covert but also leaked operations. Finally, advances in real-world AI (machine learning) might be behind a revived interest in or new sense of credibility around the actual instantiation of powerful AIs – and with them concerns about the general accident of a 'hostile' alien singularity (conscious or not; I refer here to hostility metaphorically) wiping us all out. These concerns cross-list with other titles already named in earlier categories.

(vi) Horrible humans

This label designates that form of anti-computing that doesn't 'blame' computers for societal ills but excoriates humans for taking advantage of what computers increasingly enable humans to 'get away with'. Originally it included concerns chiefly focusing on 'people' rather than governments or corporations. This capacious category now expands to accommodate rising dissent and anger about new media organizations (the platforms, chiefly) and their role in the rise of new forms of populism, extremism, intolerance, bigotry, hatred, and aggression in public spheres. Something striking about this form of anti-computing today, perhaps what makes it distinctively of the moment, is the degree to which questions concerning technology, morality, and ethics (e.g. of good and evil) are invoked at many scales simultaneously, and in relation both to personal lives and the conduct of politics and public life in general.

Moral panics (Cohen, 2011), originally defined in 1972 around media content, forms and formats, were of course always political (in that they were ideologically framed, related to dominant ideas and conceptions, and also to material conditions/relations). What is now evident is a form of anti-computing sentiment that amounts to a moral panic about the *relationship between* the personal and political; what is being critiqued is the (populist) form of the political, of politics, or political life, or public life, or civil society, that societies are (felt to be) letting happen by failing to rein in or control the computational and computational corporations. Amongst the books taking this kind of line is Sarah Jeong's *Internet of Garbage* (2018), which asks how we filter wanted content from the 'garbage' found online: 'Content platforms and social media networks do not have the power to restrain stalkers, end intimate partner violence, eliminate child abuse, or stop street harassment ...' This both identifies the problem and suggests cures, claiming that it would be possible to 'cultivate better interactions and better discourse, through thoughtful architecture, active moderation and community management'. Jeong's work is a call for responsibility to be taken up, but explicitly not a call for more radical societal change.

Jaron Lanier, an early Silicon Valley apostate, excoriates the 'designed in' toxicity of social media and suggests how to get off it. His *Ten Arguments for Deleting Your Social Media Accounts Right Now* (2018) comes with a blurb striking for its affective language; the cruel, dangerous, fearful, isolated, tribal, are all invoked:

> Social media is making us sadder, angrier, less empathetic, more fearful, more isolated and more tribal. In recent months it has become horribly clear that social media is not bringing us together – it is tearing us apart ... Jaron Lanier draws on his insider's expertise to explain precisely how social media works – by deploying constant surveillance and subconscious manipulation of its users – and why its cruel and dangerous effects are at the heart of its current business model and design.

Lanier is clear that if these are horrible human emotions, they are produced by our dangerous relationship with unethically designed machines and communication models. He goes on to argue that we need to act to help ourselves to get out of this. The 'ripping apart' effects of digital media are said to be reorganizing politics and encouraging forms of dangerous populism. An account of this is given in Mike Wendling's *Alt Right: From 4chan to the White House* (2018). This 'reveals the role of technological utopians, reactionary philosophers, the notorious 4chan and 8chan bulletin boards, and a range of bloggers, vloggers and tweeters, along with the extreme ideas which underpin the movement's thought'. Other publications look at trolling as a human behaviour generated by the global techno-social culture we live in. Notably, Whitney Phillips's

This Is Why We Can't Have Nice Things: Mapping the Relationship between Online Trolling and Mainstream Culture promises to frame trolling in classically horrible human terms, and as mainstream:

> Why the troll problem is actually a culture problem: how online trolling fits comfortably within today's media landscape. Trolling may be obscene, but, Phillips argues, it isn't all that deviant. Trolls' actions are born of and fuelled by culturally sanctioned impulses – which are just as damaging as the trolls' most disruptive behaviors. We don't just have a trolling problem, Phillips argues; we have a culture problem. This ... isn't only about trolls; it's about a culture in which trolls thrive.

Ginger Gorman's *Troll Hunting* (2019) is also insistent that trolling is widespread, but apparently discerns no coherent logic behind it. The argument, we are told, is that

> Syndicates of highly organised predator trolls systematically set out to disrupt and disturb. Some want to highlight the media's alleged left-wing bias, some want to bring down capitalism and others simply want to have some fun, even if it means destroying the victim's emotional and financial life.

Trolling produces accusatory responses attacking the forms of computational culture that enable or encourage these kinds of action. What is also evident is the desire to offer cures rather than societal-wide 'solutions' (political responses narrowly defined). Self-help is thus very prominent in the suggested readings. An indicative version comes from Sherri Mabry Gordon (2018), who offers teens help in *Coping with Online Flaming and Trolling*, and the 'Shock. Disbelief. Pain. Embarrassment' it causes.

There is a line of critique that fits into the category of horrible humans because, whilst attacking the culture of the internet as sexist, racist, and classist, it does not recognize structural inequalities. The tropes reached for here tend to be personalized – they include critiques of those who attack, and assessments of damage done to individuals. *Shame Nation* (2018), 'with a forward from Monica Lewinsky', demands the formation of new forms of internet civility, but again is unabashedly a (self-)help book: 'An essential toolkit to help everyone – from parents to teenagers to educators – take charge of their digital lives', we are told.

From a different perspective comes *Misogyny Online: A Short (and Brutish) History* (2016) from Emma A. Jane. This 'explores the worldwide phenomenon of gendered cyberhate as a significant discourse which has been overlooked and marginalised. The rapid growth of the internet has led to numerous opportunities and benefits; however the architecture of the cybersphere offers users unprecedented opportunities to engage in hate speech.' Finally, also clearly hovering between this category and one more overtly political and social, but again placed in here since the tone is one of personal distress and the mode of address is to 'you the user', is Zoe Quinn's account of gamergate and after in *Crash Override* (2017). This promises an 'up-close look inside the controversy, threats, and social and cultural battles that started in the far corners of the internet and have since permeated our online lives ... Quinn provides a human look at the ways the internet impacts our lives and culture, along with practical advice for keeping yourself and others safe online.'

(vii) Standardization/quantification

Developments in this area are critiques of machine learning and what it leads to. Of particular note is automated bias – when machine learning learns from human sources and the results are forms of accelerated or intensified, and apparently more deeply inscribed, or reinstitutionalized bias. One of the best-known examples is Tay, the bot that learned to be a Nazi. Offerings here included *Automating Inequality: How High-Tech Tools Profile, Police, and Punish the Poor*, by Virginia Eubanks (2018), and Safiya Noble's *Algorithms of Oppression: How Search Engines Reinforce Racism* (2018). Attacking in the same vein, but exploring gender bias, is Caroline Criado Perez, with *Invisible Women: Exposing Data Bias in a World Designed for Men* (2019l; see also Sara Wachter-Boettcher, 2018).

Somewhat differently, there is Cathy O'Neil's *Weapons of Math Destruction* (2017), in which a 'Wall Street quant sounds an alarm on the mathematical models that pervade modern life – and threaten to rip apart our social fabric':

> We live in the age of the algorithm ... this should lead to greater fairness ... And yet, as Cathy O'Neil reveals in this urgent and necessary book, the opposite is true. The models being used today are opaque, unregulated, and incontestable, even when they're wrong. Most troubling, they reinforce discrimination.

O'Neil's sense of new architectures of unfreedom – of structural developments not amenable to personal solutions, but addressable only by way of some more substantial series of changes, resonates with Morozov's *To Save Everything, Click Here: Technology, Solutionism, and the Urge to Fix Problems that Don't Exist*. Published in 2014, this continues to arise as a contemporary reference point in reframing and refinding earlier critiques of standardization/solutionism and reinvoking them in relation to data and discourses of 'smart'. As the blurb tells us: 'Our gadgets are getting smarter … we're told … it will even make public life – from how we're governed to how we record crime – better. But can the digital age fix everything? Should it? By quantifying our behaviour, Evgeny Morozov argues, we are profoundly reshaping society – and risk losing the opacity and imperfection that make us human.'

(viii) Too much information

The final category in the taxonomy found space for works responding to reviving concerns around over information overload, but also included anxieties about the fetishization of information capture, the fetishism of facts, what is viewed as the overproduction of information – whether as data, text, image – and the prioritization of production and circulation over interpretation. Older terms expressing this concern included fears around data deluge, or information overload. There is also the now largely forgotten term 'information anxiety', credited to Richard Saul Wurman who wrote a book of that name (1989). Other early reference points include 'As We May Think' (Bush, 1945), and/or Alvin Toffler's 1970s best-seller, *Future Shock*.

Today, slightly fading now but still thrown up as a reader suggestion, is Nicholas Carr's *The Shallows* (2011): 'not since Gutenberg invented printing has humanity been exposed to such a mind-altering technology. *The Shallows* draws on the latest research to show that the Net is literally re-wiring our brains inducing … superficial understanding …'. These concerns map onto the attention economy and the demands it makes, and underscore ways that connections between this category and others are tighter than they were; the use of automation to handle information that is 'too much' for humans, and attendant concerns around autonomy are of note here.

Where this category has re-emerged is around critiques of information politics – the hollowing-out of reflection via always-on news, the prioritization of the virtual – as a mode of social being, the fragmentation of earlier forms of public life (public spheres, commons), and the rise of balkanization. An inaugural trope in this category concerns filtering, and its deleterious effects continue to be observed; the move is from zip codes to filter bubbles and echo chambers. Pariser's *Filter Bubble, What the Internet Is Hiding from You* (2012), an early swallow in a big summer, is still invoked. *The Filter Bubble* was part of that rising tide of dismay that later reached huge proportions around fake news, Trump's election, and (in the UK) Brexit. The point to be made here is that information overload surfaces with a new kind of slant; now the worry is at least as much about the various 'cures' – attention, selection, automated selection, changes in reading habits towards scanning (sixty-four books, for instance …), nudge, filtered news, and the rest – as about the sheer amount of information. This at least is what surfaces through the Amazon trawl which threw up these latter points, rather than evidencing a concern with information and/as a deluge per se. These concerns do not necessarily link to big data and solutionism, nor to concerns around information or knowledge as *complete* but, rather, focus on how the flood of information challenges interpretation and raises questions about explainability.

A conclusion

This exercise in distant reading suggests to me that the original taxonomic division of anti-computing, which set out a series of persistent characteristics, does connect to (and does enable us to grapple with and reframe) anti-computing as it arises today. It is justifiable to assert that anti-computing persists across the decades of computational instantiation in recognizable forms, even as it also shifts, morphs, finds new targets, or reshapes older ones. It also changes. The long-standing sub-categories continue to make sense but they strain and bulge, are deformed and reformed as they take in contemporary anti-computing formations. These deformations articulate the form anti-computing takes in the specific moment, in relation to earlier forms, in relation to the time in which it finds itself, in relation to technological developments of that time, and in relation to its political culture.

Chapter 1 also set out some supplementary taxonomies for anti-computing, cross-cutting the main series. One categorized anti-computing strains by dividing various forms, degrees, and kinds of affective engagement to produce an emotional register (see Figure 1). This offers a key to understanding the contemporary formation. The Amazon crawl points to the intensity of responses to the computational currently evident, and the degree to which these continue both to distinguish and to blur distinctions between concerns around the ontology of the computational and those related to issues of social and political power, control, and domination. Anger around gender and race bias and the concern about human hatred is palpable as a framing of many of the works found in the Amazon trail (and clear elsewhere, of course). The strength of the feeling of disgust, not only with the computational as machine culture but with computational culture as it has mirrored the dark side of human culture, is marked. Supplementing this, a more ambivalent orientation – and often one with a less intense affective charge – is evident in the framing of works exploring singularity and/as AI, and the issue of belief and the matter of its relation to *caring* here raises its head again. Chapter 7 grappled with the paradoxical mix of formal acceptance and emotional and even rational scepticism surrounding singularity's claims. Two emotions – viscerally felt disgust, and an insincere fatalism (rather than a cruel optimism, perhaps) arising out of a form of unbelief – are both in evidence in contemporary anti-computing's affective geography, and their measure as it were is reallocated across the various categories of the main taxonomy. There is remarkable cheer in many of the publications declaring the end of the world, and the help books are (not surprisingly, given their sales imperative) far more willing to countenance 'solutions' than their own arguments would appear to suggest is possible.

Algorithms of course do not feel emotion. Moreover, they do not necessarily parse human emotions well. Perhaps it was the emotional inadequacy of Amazon's algorithm that meant that every so often it threw up a blisteringly positive account mixed in with the negative analysis, hostility, dissent, and relentless advice (amongst the celebrations from the Left was Bastani's 2020 *Luxury Automated Communism*). But this can also be understood as a symptom of the entanglement between these two orientations, pro- and anti-computing, pointing to their intertwined histories, and to the roles each plays in the future of the other, and this seems to me a sustainable view, given the overarching formations and the particular cases explored in this book. Anti- and pro-computing formations, that is, travel *on* together, each being less visible, or powerful at various times, in relation to the other, even whilst the systematic dominance of the one over the other persists – and has persisted across the decades of the development of computational capitalism, with its relentless prioritization of more growth.

These two sets of observations, on remapped affectivity, and on the entanglement between computational avidity and dissent, inform some concluding remarks. These include a moment of critique and dissent. What I dissent from is disgust and morality as sufficient frames through which to critique the computational, and computational culture in general, in neoliberal times. I also distance myself from self-help (or how to feel *better*) as a response to computational ills. These forms of anti-computing are, if not encouraged, then given some purchase in current times even by the industry itself, notably by the platforms, for instance. To contest them is to undertake a form of conditional anti-computing which is at times *anti*-anti-computing. It demands a commitment to radical structural politics rather than personal solutionism, and/or the advocacy of automated cures to computer-assisted human badness (behaviourism at one end, bot-censorship at the other; the age of the super-nudge).

Elaborating this demands back-pedalling slightly to regloss the claim that the anti-computational comes with the computational. Consider that forms of absolute refusal, or the desire to end computing absolutely, are either vanishingly rare or have existential implications. When I began this book I intended to write a chapter on total refusal but it turned out to be almost impossible to find. Even preppers connect, albeit carefully; religious movements against Western modernity, notoriously, use the internet; whilst primitivism as a movement is even, if not conspicuously, not entirely against computer use (see e.g. Zerzan, 2012). Seeking to abolish the computer itself as a mode of production, or the computational as a form of life, or even to abandon all forms of computational thinking, is rarely what is called for, or desired. To some extent this is a matter of practicalities; in an era of intense saturation, where there is no way out, how low can you go? Or, rather, where can you go? Or how low and where can you go in a globalized world encircled with satellites, where older forms of communication shrivelled up and died in the face of the new? One of the arguments of this book has been about forgetfulness and the appearance of naturalness in the technological which becomes an embedded part of life. If this constrains thinking within the horizon of the computational, it is also because life itself is lived within this horizon.

There is one site where complete decomputerization is countenanced. Certain visions of the end of computing link in to accounts of the end of the human, the end of human civilization, perhaps the end of the Earth. A form of anti-computing shades into apocalypse-thinking, and in this way also perhaps into some of the most anti-human wings of environmental politics; the presumption that the Earth is better off, or only *can* survive and regreen without us, and without any of our technologies. Derrida (1984) argued that this horizon is unthinkable – even speculatively – and this side of the apocalyptic prediction/desire for apocalypse, anti-computing is less than absolute. It variously seeks a kind of amelioration, or a process of modification, or a new deal. It wants to check impulses, or companies, or expansion into certain areas of life. That doesn't make it altogether – or even necessarily – reformist. The anti-computational may constitute an element of radical, critical, even revolutionary thinking on the technological. Indeed, because anti-computational is essential to understanding the computational, is part of computational culture, it can be used to define and understand another set of divisions within contemporary techno-cultural thinking; that is the divide between anti-computing positions that entail a radical (structural) critique of computational capitalism and those that do not. Here I myself come off any fence I might be thought to still be sitting on and declare for systematic critique.

In case I am misunderstood here, that doesn't mean arguing that real interest, or significance, in the end inheres only in writings or activities exploring radical or systematic critiques of the technological, or computational, moment. On the contrary, exploring the range of forms anti-computing has taken and takes, as intellectual points of view, shared positions, public arguments, as discourses that emerge, submerge, revive, and appear in new registers (personal, political, moral, political, as campaign, as life story, as political event, as software, as fiction) as they travel, and looking at how they travel and how they land has been the point of this book. Understanding these positions and their operations can contribute to understanding and can assist in breaking through that presentism and those forms of 'realism' that constrain thinking about the technological. I put this in scare quotes with reference to Le Guin's discussion of more radical realism invoked in Chapter 7, and also with reference to Mark Fisher's (2009) elaboration of capitalist realism as that horizon which appears to prescribe the possible.

Amongst these constraints are those that systematically drive thinking towards the personal and/or towards a particular conception of responsibilization and atomization; social being as individuated being, so that the anti-computational is itself experienced as a personal response and one that responds to the discrete issue rather than a historically instantiated and developing world order; a techno-political economy.

This orientation explains the heavy leaning towards help, the personalization of these modes of address, which are a characteristic of the anti-computing books we are recommending to each other through our Amazon-circulated 'reading' practices. They are, of course, at one with computational shifts in social life, social commons, and social being. We are increasingly invited, indeed organized, in so far as possible directed/nudged, into positions in which our sense of where we may respond, our sense of 'responsibility' even, and our sense of where the solution to our problem may be found, shifts towards the personal/personalized sphere.

Questioning the implications of this shift in focus, *as itself a mode of critical anti-computing*, rather than simply locating an anti-computing trope *within* it, is therefore significant. Anti-computational thinking on the whole breaks with the assumption that if it is bad today it will get better tomorrow. It can help to constitute an effective intervention partly because it does not partake in a particular kind of expectation for the automated revivification of faith in progress through technology, the kind of technological optimism that says technology will win through, will provide the cure, will make everything new again. Instead, it reaches for those other stories, those other tropes and traditions, through which it thinks, across which it invokes or makes an understanding of the new in the context of the pain of the old, and its trouble.

This can produce a nostalgic orientation, and one that is, in the current climate, in danger of a kind of revived parochialism, certainly as it relates to global culture on the internet, beyond the walls many in control want to build. But that other kind of anti-computing can find in this refusal to be forgetful a form of hope, the kind of hope Walter Benjamin found in refusing to let the past – and its costs – be covered over; that sees that it rises, and sometimes helps it to do so (Benjamin, 2006). Perhaps we should follow Terry Eagleton (2015) in distinguishing between hope (which contains possibility and uncertainty) and optimism (which gambles on less, in exchange for more security). If there is a sense of cruel *pessimism* (contra Berlant's [2011] well-known formulation of cruel optimism) in the fatalistic anti-computational assessments of the place we find ourselves in, there is also the hope that comes with the demands anger makes. Some of the responses to terrorist events involving networks and social media, to algorithmic bias, to the alt right and its channels, those responses that demand solidarity and refuse hatred and that demand an exploration of what structured events rather than simply asking *how* they circulated, exhibit that hope. Let that be a place to end.

Bibliography

Aiken, Mary. 2017. *The Cyber Effect*. London: Hachette.
Alter, Adam. 2017. Irresistible: The Rise of Addictive Technology and the Business of Keeping Us Hooked. London: Bodley Head.
Barlow, John Perry. 1994. 'Stopping the Information Railroad', keynote address, Winter USENIX Conference, San Francisco, 17 January.
Barrat, James. 2013. *Our Final Invention*. London: Thomas Dunne.
Bartlett, Jamie. 2015. *The Dark Net: Inside the Digital Underworld*. London: Heinemann.
Bartlett, Jamie. 2018. *The People vs Tech: How the Internet is Killing Democracy (and How We Save It)*. London: Ebury Press.
Bassett, Caroline and Ben Roberts. 2020. 'Automation Now and Then: Automation Fevers, Anxieties and Utopias', *New Formations*, 98: 9–28.
Bastani, Aaron. 2020. *Fully Automated Luxury Communism: A Manifesto*. London: Verso
Benjamin, Walter. 2006. 'On the Concept of History', in *Selected Writings, 4: 1938–1940*. London: Harvard University Press.
Berlant, Lauren. 2011. *Cruel Optimism*. London: Duke University Press.

Broussard, Meredith. 2018. *Artificial Unintelligence*. London: MIT Press.
Bush, Vannevar. 1945. 'As We May Think', www.theatlantic.com/doc/194507/bush.
Carr, Nicholas. 2011. *The Shallows. How the Internet Is Changing the Way We Think, Read, and Remember*. London: Norton.
Carr, Nicholas. 2016. *The Glass Cage: Where Automation is Taking Us*. London: Norton.
Chace, Callum. 2018. *Surviving AI* (2nd edn). London: Three Cs.
Christian, Brian, and Tom Griffiths. 2017. *Algorithms to Live by: The Computer Science of Human Decisions*. London: Collins.
Cohen, Stanley. 2011. *Folk Devils and Moral Panics: The Creation of the Mods and Rockers* (3rd edn). London: Routledge.
Eagleton, Terry. 2015. *Hope without Optimism*. New Haven: Yale University Press.
Eubanks, Virginia. 2018. *Automating Inequality: How High-tech Tools Profile, Police, and Punish the Poor*. New York: St Martin's Press.
Fisher, Mark. 2009. *Capitalist Realism: Is There No Alternative?* London: Zero Books.
Foer, Franklin. 2018. *World Without Mind: The Existential Threat of Big Tech*. London: Penguin.
Ford, Martin. 2016. *Rise of the Robots*. London: Basic Books.
Fry, Hannah. 2019. *Hello World: How to Be Human in the Age of the Machine*, New York: Norton.
Galloway, Scott. 2017. *The Four: The Hidden DNA of Amazon, Apple, Facebook and Google*. London: Random House.
Gannon, Emma. 2016. *Ctrl, Alt; Delete: How I Grew Up Online*. London: Ebury Press.
Gazzaley, Adam and Larry Rosen. 2016. *The Distracted Mind: Ancient Brains in a High Tech World*. London: MIT Press.
Gilroy Ware, Marcus. 2017. *Filling the Void: Emotion, Capitalism and Social Media*. London: Duncan Baird Publishers.
Goodin, Tanya. 2017. *Time To Log Off: Your Digital Detox for a Better Life*. London: Hachette.
Goodman, Marc. 2015. *Future Crimes: Inside The Digital Underground and the Battle for Our Connected World*. London: Penguin.
Gordon, Sherri Mabry. 2018. *Coping with Online Flaming and Trolling*. New York: Rosen.
Gorman, Ginger. 2019. *Troll Hunting: Inside the World of Online Hate and Its Human Fallout*. London: Hardy Grant.
Graeber, David. 2015. *The Utopia of Rules: On Technology, Stupidity, and the Secret Joys of Bureaucracy*. London: Melville House.
Gregg, Melissa. 2011. *Work's Intimacy*, Cambridge: Polity.
Harris, Michael. 2018. *Solitude, in Pursuit of a Singular Life in a Crowded World*. London: Penguin.
Harrison, Emma Elizabeth. 2020. 'Activism, Refusal, Expertise: Responses to Digital Ubiquity'. Doctoral thesis, University of Sussex.

Henderson, Lance. (2017–18) *Tor and the Deep Web*. Independently published.
Hern, Alex. 2019. 'New AI Fake Text Generator May Be too Dangerous to Release, Say Creators', *The Guardian*, 14 February.
Hunter, K.N. 2017. *How Do I Live Without Trolling You: A Look into the LeAnn Rimes Hate Culture Online*. Amazon.com services.
Illing, Sean. 2018. 'Interview with Sherry Turkle', *Vox*, 27 March.
Jane, Emma A. 2016. *Misogyny Online: A Short (and Brutish) History* London: Sage.
Jeong, Sarah. 2018. *The Internet of Garbage*. New York: Vox Media.
Karppi, Tero. 2018. *Disconnect: Facebook's Affective Bonds*. Minneapolis: University of Minnesota Press.
Kasket, Elaine. 2019. *All the Ghosts in the Machine: The Digital Afterlife of Your Personal Data*. London: Robinson.
Keen, Andrew. 2008. *The Cult of the Amateur* (rev. edn). London: Nicholas Brealey Publishing.
Keen, Andrew. 2019. *How to Fix the Future: Staying Human in the Digital Age*. London: Atlantic Books.
Lanier, Jaron. 2018. *Ten Arguments for Deleting Your Social Media Accounts Right Now*. Oxford: Bodley Head.
Leonhard, Gerd. 2016. *Technology vs. Humanity: The Coming Clash between Man and Machine*. New York: Fast Future Publishing.
Mantilla, Karla. 2015. *Gendertrolling: How Misogyny Went Viral*. London: Praeger.
Marchant, Danielle. 2018. *Pause: How to Press Pause Before Life Does It for You*. London: Aster.
Masko, Dave. 2019. *Tech 5G Networks Nervous System Threats 2019: Donald Trump 5G Erodes Global Order, Singularity Out-Of-Control*. Kindle: Dave Masko Copyright.
Mason, Paul. 2019. *Clear Bright Future*. London: Penguin.
Masulo Chen, Gina. 2017. *Nasty Talk: Online Civility and Public Debate*. London: Palgrave Macmillan.
McNamee, Roger. 2019. Zucked: Waking Up to the Facebook Catastrophe. London: HarperCollins.
Miller, Toby and Richard Maxwell. 2020. *How Green Is Your Smartphone?* London: Polity.
Moore, Martin. 2018. *Democracy Hacked: Political Turmoil and Information Warfare in the Digital Age*. London: One World Publications.
Moretti, Franco. 2013. *Distant Reading*. London: Verso.
Morozov, Evgeny. 2012. *The Net Delusion: How Not to Liberate the World*. London: Penguin.
Morozov, Evgeny. 2014. *To Save Everything, Click Here: Technology, Solutionism, and the Urge to Fix Problems that Don't Exist*. London: Penguin.
Nagle, Angela. 2017. *Kill All Normies: Online Culture Wars from 4chan and Tumblr to Trump and the Alt-right*. London: Zero Books.

Noble, Safiya Umoja. 2018. *Algorithms of Oppression*. New York: New York University Press.
O'Neil, Cathy. 2017. *Weapons of Math Destruction: How Big Data Increases Inequality and Threatens Democracy*. London: Penguin.
Pariser, Eli. 2012. *The Filter Bubble: What the Internet Is Hiding from You*. London: Penguin.
Pasquale, Frank. 2015. *The Black Box Society: The Secret Algorithms that Control Money and Information*. Cambridge, MA: Harvard University Press.
Perez, Caroline Criado. 2019. *Invisible Women: Exposing Data Bias in a World Designed for Men*. London: Chatto and Windus.
Phillips, Whitney. 2016. *This Is Why We Can't Have Nice Things: Mapping the Relationship between Online Trolling and Mainstream Culture*. London: MIT Press.
Quinn, Zoe. 2017. *Crash Override: How Gamergate (Nearly) Destroyed My Life, and How We Can Win the Fight Against Online Hate*. New York: Public Affairs Publisher.
Reagle, Joseph. 2015. *Reading the Comments: Likers, Haters, and Manipulators at the Bottom of the Web*. London: MIT Press.
Rushkoff, Douglas. 2019. *Team Human*. London: Norton.
Scheff, Sue. 2018. *Shame Nation*. London: Sourcebooks.
Scott, Laurence. 2015. *The Four-Dimensional Human: Ways of Being in the Digital World*. London: William Heinemann.
Senker, Cath. 2017. *Cybercrime and the Darknet*. London: Arcturus Publishing.
Seymour, Richard. 2019. *The Twittering Machine*. London: Indigo Press.
Solnit, Rebecca. 1995. 'The Garden of Merging Paths', in J. Brook and Iain Boal (eds) *Resisting the Virtual Life, the Culture and Politics of information*. San Francisco: City Lights, 221–234
Stallabrass, Julian. 1996. *Gargantua*. London: Verso.
Storr, Will. 2018. *Selfie: How the West Became Self-obsessed*. London: Picador.
Taplin, Jonathan. 2017. *Move Fast and Break Things: How Facebook, Google and Amazon Have Cornered Culture and Undermined Democracy*. London: Macmillan.
Toffler, Alvin. 1970. *Future Shock*. London: Penguin.
Turkle, Sherry. 2005. *The Second Self: Computers and the Human Spirit*. London: MIT Press.
Turkle, Sherry. 2011. *Life on the Screen*. London: Simon and Schuster.
Turkle, Sherry. 2016. *Reclaiming Conversation: The Power of Talk in a Digital Age*. London: MIT Press.
Turkle, Sherry. 2017. *Alone Together: Why We Expect More from Technology and Less from Each Other* (3rd edn). London: Basic Books.
Vigna, Paul and Michael J. Casey 2016. *Cryptocurrency: How Bitcoin and Digital Money Are Challenging the Global Economic Order*. London: Bodley Head.

Wachter-Boettcher, Sara. 2018. *Technically Wrong: Sexist Apps, Biased Algorithms, and Other Threats*. London: Norton.
Wendling, Mike. 2018. *Alt-Right: From 4chan to the White House*. London: Pluto Press.
Wiener, Norbert. 1961. *Cybernetics, or Control and Communication in the Animal and the Machine*. Cambridge, MA: MIT Press.
Wu, Tim. 2017. *The Attention Merchants: The Epic Struggle to Get inside Our Heads*. London: Atlantic Books.
Wurman, R.S. 1989. *Information Anxiety*. New York: Doubleday.
Zerzan, John. 2012. *Future Primitive Revisited*, Port Townsend: Feral House.
Zuboff, Shoshana. 2019. *The Age of Surveillance Capitalism: The Fight for a Human Future at the New Frontier of Power*. London: Profile Books.

Index

Note: literary works can be found under authors' names. 'n.' after a page reference indicates the number of a note on that page.

Ad Hoc Committee on the Triple Revolution 105–106, 110, 112–113
  Report on the Triple Revolution 102n.66, 105–107, 110–113, 115, 131n.8
AFL/CIO 117
AI *see* artificial intelligence
Aiken, Mary, *The Cyber Effect* 225
Alexa 182
algorithms 5, 9, 22, 39, 54, 141, 231, 234, 238
  Amazon and 218–219, 223, 234
Alter, Adam, *Irresistible* 225
Amazon 3, 217–237 *passim*
Amigoni, David 155
Amis, Kingsley 161
Anthony, Scott 50
anti-computational thinking 23, 169, 219, 221, 237
Apple 39, 93, 219
archaeology (of knowledge) 8, 43–45, 47
  see also media archaeology
Arendt, Hannah 11, 105, 107–108, 116, 118–130 *passim*
  Conference of the Congress of Scientists on Survival 106
  Human Condition, The 11, 107–108, 121, 123–124, 126–128
  Little Rock 122
  On The Human Condition (conference paper) 11, 105, 107–108, 118, 127–128
  On Violence 128
  on technology 108, 123–124, 126–128
Armer, Paul 114
Arnold, Matthew 142
artificial intelligence 4–5, 20–22, 36, 56, 168–182 *passim*, 186–206 *passim*, 222–224, 228, 234
  OpenAI GPT2/3 language model 220
automation 5, 11, 20–21, 39, 63n.5, 69–74 *passim*, 105–130 *passim*, 145, 162, 223, 232
anxiety 2–3, 34, 72, 217, 223; *see also* digital anxiety; 'information anxiety'
computational 11; *see also* cybernation
computer-delivered 20; *see also* cybernation

automation (cont.)
  scare 108, 129; *see also* cybernation scare
  see also database

Babbage, Charles 9
  engine 72
Bachelard, Gaston 43
Bagrit, Leon, *The Age of Automation* 110
Baker, Kenneth 92
Barad, Karen 29, 60
Barbrook, Richard 40
Barlow, John Perry 225
Barrat, James, *Our Final Invention* 223
Bartlett, Jamie, *The People versus Tech* 217, 222
Bassett, Caroline 28–29, 170, 178–180, 192, 217
  AI and Society 170
Bastani, Aaron, *Luxury Automated Communism* 234
BBC 41, 84
  Reith Lecture Series 110
Beck, Ulrich
  'risk society' 22
behaviourism 170–171, 178–179, 183n.3, 235
Bell, Daniel 129
Benhabib, Seyla, *Politics in Dark Times* 124
Benjamin, Walter 38, 45, 62, 195, 238
Benn, Anthony Wedgwood 98n.5
Bentham, Jeremy 126
Benthamism 125, 147, 154–155, 157
  technologico- 5, 151–152, 154–156, 158, 163, 224
Bentley, Elizabeth 77–78
Berenyi, Peter 69
Berkeley, Edmund 36, 224
Berlant, Lauren 238

Berry, David 9
Bey, Hakim 8
big data 3, 23, 35, 126, 176, 203, 233
Bill on Right of Privacy, UK (1969) 92
Birmingham School of Cultural Studies 53–54, 162
Blackford, Russell 188, 200
Blas, Zac 26
Block, James E., 'The Selling of the Productivity Crisis' 129
Boggs, Grace 115, 132n.15, 133n.25
Boggs, James 113, 115–118, 121, 131n.6, 132n.15
Bogost, Ian 27
Borges, Jorge Luis 16, 218
Bould, Mark 192
Bradbury, Malcolm 161
Braybrooke, Kat 40
Brexit 140–141, 163, 220, 233
Broussard, Meredith 224
Brozen, Yale 112, 131n.9
Budenz, Louis 77–78
Bush, Vannevar, 'As We May Think' 232
Butler, Judith 61

Cambridge Analytica 220
Cameron, Angus 81
Canovan, Margaret 123–124
capitalism 13, 19, 24, 38, 58–59, 73, 94, 115, 209
  computational 3, 8, 11, 35, 55, 57–58, 203, 208, 235–236
  informational 2, 222
  surveillance 222
Carr, Nicholas
  Glass Cage, The 223
  Shallows, The 232
Centre for the Study of Democratic Institutions 109

Chase, Callum, *Surviving AI* 223
Chomsky, Noam 179, 183n.11
Christchurch massacre 34–35, 238
Chun, Wendy 48
cloud, the 5, 10
'code and language' dilemma 46–48, 206
Cohen, Ronald D. 77
Cohn, Roy 79, 82, 99n.28
Cold War 41, 70–71, 80, 85, 95–96, 106, 147
Collini, Stefan 148
communism 73, 78–79, 217, 234
communist 70, 78–82, 84, 98n.13, 99, 99n.23
　anti- 70, 76–77, 79, 82, 96
　Party 69–70, 75–76, 114
computational
　culture 2, 8–9, 11, 16, 21, 28, 59, 217–218, 230, 234–236
　thinking 9, 169, 219, 235
　see also anti-computational thinking
Computers and People magazine 36
Conference on the Cybercultural Revolution, First Annual 105–107, 112, 116, 122, 128–130
The Evolving Society 107
counter-culture 5, 36, 70–71, 83–97 *passim*, 160
Crick, Francis 161
cultural studies 45, 51–55, 144, 162
　digital 196
　see also Birmingham School of Cultural Studies
Cvetic, Matt 77–78, 99n.15
cybercultural
　era 121
　idealists 5
　revolution 105–107, 110, 112, 117, 128, 133n.23
　see also Conference on the Cybercultural Revolution
　society 18
cyberculture 107, 111, 113, 116, 129–130, 132n.23
cybernation 2, 11, 63n.5, 105–121, 123, 126, 128–130, 131n.9
　revolution 106, 111, 129
　scare 105, 107–108, 129–130; see also automation scare
　see also automation
cybernetics 9, 41, 106, 108–109, 128, 131n.4, 173
cyberspace 130, 132n.23, 187, 195–197, 201–202, 207

Daily Mirror, The 85–87, 98
　see also Ward, Chris
database 5, 19, 21, 58, 69–97 *passim*, 180
　anxiety 69, 71, 84, 89
　automation 71–72
　form OHID see under Matusow, Harvey
　society 69, 71, 73, 93, 102n.77
data capture 3, 12, 93, 220, 225
　see also information capture
Dean, Jodi 72
Deep Web, the 225
Delamont, Sarah 161–162
Deleuzian entanglement 189
del Val, Jaime, and Stefan Lorenz Sorgner, 'A Metahumanist Manifesto' 189–190
Derrida, Jacques 183n.9, 236
detox, digital see digital
digital
　age 232
　anxiety 218
　bio- 20, 203
　culture 5, 163, 224
　detox 21, 224, 226

digital (cont.)
  humanities 7, 51–52, 55, 143, 193
    see also cultural studies
  media 229
    archaeology 45, 51
    history 41
    platforms 225
    research 42
  memorialization 93
  post- 11, 35–36, 204
  surveillance 98n.1
digital, the 2–3, 5–6, 11, 19, 57, 189, 203, 219, 222, 224
digitalization 73–74
Dizikes, Peter 146–147
DOCTOR (script) 168, 178
Douglas, Gordon, *I was a Communist for the FBI* 99n.15
Durham Peters, John 69, 78, 96–97

Eagleton, Terry 238
Edwards, Paul 39, 63
Eliot, T.S. 158
ELIZA (script) 54, 168–170, 172, 175–183
Ellison, Harlan, *Dangerous Visions* 112
Ellul, Jacques 45, 170
Elsaesser, Thomas 45, 72
Ernst, Wolfgang 46–51, 54–55, 76, 98n.13
Eubanks, Virginia, *Automating Inequality* 231

Facebook 98n.8, 183, 225–226
fake news 35, 220, 227, 233
fantasy (genre) 192, 202, 204, 212
  upload 187–188, 195, 208–212
fascism 69, 72, 84, 147
Fazi, Beatrice 56, 183n.5
FBI 76–77, 79, 81, 99n.15

feminism 46, 54, 61, 105–106, 113, 129, 144, 161–162, 197, 201, 213n.9, 220, 227
  cyber- 40
Ferrando, Francesca 189–190, 206, 212n.2
Existenz 189
filter bubble 141, 233
Fisher, Mark 212, 237
Flanders and Swann 141, 149–152, 159–162
  At the Drop of Another Hat 141, 149, 151, 159–160, 164n.15
  'Thermodynamics Song' 141–142, 149–150, 159–160
Foer, Franklin, *World without Mind* 222
Ford, Martin, *Rise of the Robots* 223
forgetfulness, technological 55–59, 119, 132n.16, 236
Foucault, Michel 7–8, 16, 34, 43–49 *passim*, 54, 56, 60–62, 154–155, 209
  Archaeology of Knowledge 43–44
  Discipline and Punish 154
  Power/Knowledge 44–45
Frankfurt School 20
Franklin, Rosalind 161
Freccero, Carla 61
Friedan, Betty 113, 117, 121, 131n.10
Fuller, Matt 58

Galbraith, John Kenneth, *The Affluent Society* 111
Galloway, Scott, *The Four* 219
Ganz, Samuel 129, 132n.22
Gates, Bill 193
Gebru, Timnit 7
Getts, Clark H. 79, 99n.22

Index

Gibson, William 130, 132n.23
   Neuromancer 196–197
Gitelman, Lisa 40–41
Godard, Jean-Luc 75, 98n.12
Gogol, Nikolai 208
Goldberg, Maxwell 114, 131n.9
Goodin, Tanya, *Time to Log Off* 226
Goodman, Edith 118
'goodness, automatic' 178–179
   see also Skinner, Burrhus F.
Google 30, 39, 219, 225
Gordan, Lewis R. 11
Gordon, Sherri M., *Coping with Online Flaming and Trolling* 230
Gorman, Ginger, *Troll Hunting* 230
GPT 2/3 language model 220
Graeber, David 209, 227

Haddon, Leslie 42
Hall, Stuart 53–54, 108, 162
Haraway, Donna 7, 97, 194, 201, 222
Harding, Denys W. 154
Harris, Michael, *Solitude* 225
Harrison, Emma 218
Hartmann, Maren 10
Hawking, Stephen 5, 193
Hayles, N. Katherine 26, 190–191, 206
Heidegger, Martin 128, 212n.2
Hilbert, David 9
Hilton, Alice M. 106–128 *passim*
Hinze, Firer 122–123
history 7, 14, 38–47 *passim*, 55, 61–62, 97, 112, 171, 191
   anti-computing 46, 61–62
   bodies 14, 49, 60–62; *see also* Traub, V.
   computing 6, 8, 29, 34, 39–40, 63, 94–95
   media 40–42, 47, 72
   medium 42, 50, 72
Houghton, James 115

HUAC (House Un-American Activities Committee) 69–70, 72, 76–77, 79, 84, 91
Huhtamo, Erkki 17, 49, 63n.5, 110, 130
Huizinga, Johan 207
Huxley, Thomas H. 142, 159, 163n.4

IBM 41, 84, 88–90, 92–93, 106
identity 36, 40, 69–76, 84, 87, 93, 168
idleness 106, 120–121, 132n.17
Industrial Revolution 119
industrialism 224
industrialization 152–153, 155–156, 157
'information anxiety' 232
information capture 23, 232
   see also data capture
Instagram 225
Institute for Cybercultural Research 100n.33, 113
Internet of Things 9
ISADPM (International Society for the Abolition of Data Processing Machines) 70–71, 85–86, 89–90, 92
   anti-computing league 69, 72, 84, 89
Ishiguro, Hiroshi 182
IT (*International Times*) 70, 84, 89

Jameson, Frederic 13, 59, 202, 204
Jane, Emma A., *Misogyny Online* 230
Jencks, Clinton 79, 81
Jeong, Sarah, *Internet of Garbage* 229
Jobs, Steve 39
Johnson, Lyndon B. 106, 112, 131n.3
   'Great Society' 106, 131n.4
Joy, Bill 17, 193

Justice Department (US) 77, 79, 99n.26

Kafka, Franz 126, 128
Kahn, Albert E. 74, 77, 81–82, 93
  Matusow Affair, The 99n.26
Karpii, Tero, *Disconnect* 226
Keen, Andrew, *The Cult of the Amateur* 227
Kindle 219
Kittler, Friedrich 42, 46–48, 55, 206
Kress, Nancy, *Beggars in Spain* 191
Kurzweil, Ray 182, 188

labour 7, 20, 63n.5, 106–132 *passim*
  movement 110–111
  organization 113
  rights 114
  society 105
Lacey, Kate 41
Lanier, Jaron 3, 5, 9, 48, 57, 226, 229
  Ten Arguments for Deleting Your Social Media Accounts Right Now 229
  You are not a Gadget 3
Larkin, Philip 145, 159, 161
Latour, Bruno 7, 96, 194
Lazzarato, Maurizio 209–210
  Making of the Indebted Man, The 209
Leavis, F.R. 5, 140–145, 147–148, 150–163, 224
  Education and the University 152
  Nor Shall My Sword 158
  Restatement for Critics 152
  Richmond Lecture 147–148, 154–155
  Scrutiny 140, 142–143, 152–154, 156
leisure 105–130 *passim*
  society 105–129 *passim*

Le Guin, Ursula K. 186–188, 194–195, 203, 212, 237
Leith, Sam 146–147
Lenz, Claudia 120–121, 126
Leonhard, Gerd, *Technology vs. Humanity* 223
Levy, Steven 39
Lewinsky, Monica 230
liberalism 140, 142, 151
  neoliberalism 11, 24, 62
Lichtman, Robert M. 77
Light, Ann 42
Living Certificates 113
London Film Co-Op 70, 84
Loth, David 76, 98n.13
Lotus 1-2-3 12
Lovecraft, H.P. 202
Lovelace, Ada 9
Lovelock, James 202
Lovink, Geert 46–47
Luddism 5, 156
Luddite 5, 30n.1, 86, 143, 147, 155–157

Macy conferences 108
Maddox, Brenda 161
Madison, David 94
Malik, Rex 90–91, 98n.3, n.5, 102n.66
  see also New Scientist
Marchant, Danielle, *Pause* 226
Marx, Karl 43, 112, 120
Marxism 54, 173
  critiques 140
Mason, Paul, *Clear Bright Future* 218
Matthews, Sean 160
Matusow, Harvey 69–102
  'Anti-Matter' columns 85, 99n.28, 101n.45
  Beast of Business, The 86–87, 89, 91, 93, 98n.3
  Bentley, Arvilla (wife) 79, 82, 99n.24
  Cocky (website) 93

Counterattack (publisher) 77–78, 80, 82
False Witness (autobiography) 73–74, 76, 82
as Job Matusow 93
Lockwood, Anna (wife) 92
Matusow Collection (repository), University of Sussex 73–77
On Human Individual Dignity (database form OHID) 83, 88
prison 82
recantation 70, 74, 79, 81–82, 84, 94
Stringless Yoyo, The 73, 84–85, 98n.9
War between Fats and Thins 84
Wolfe (publisher) 92
Maxwell, Richard 228
McCarthy, Joseph 69–94 *passim*, 114
McCarthyism 72, 77, 80, 84, 99n.23
MacKenzie, Adrian 144
McLuhan, Marshall 45, 53, 71–72, 83, 100n.33
Medhurst, Andy 141
media
  archaeology 8, 14, 27, 42–55, 61–62, 96–97, 212
  studies 46, 53
medium theory 27, 34, 40–46 *passim*, 53, 69, 71, 77, 83
metahuman 190
metahumanism 189–190, 206
  see also del Val and Sorgner, 'A Metahumanist Manifesto'
Metropolis (film) 201
  False Maria 201
Meyrowitz, Joshua 53
Michael, Donald N. 109, 111, 113, 116
  Silent Conquest, The 109–110, 113, 116

Miéville, China 42, 187, 192, 203–207
  City and the City, The 42
  Embassytown 187, 204–207, 213n.12, n.13
  Perdido Street Station 204
Miller, Arthur 70, 92
Miller, Toby 228
Minsky, Marvin 178
MIT (Massachusetts Institute of Technology) 168, 178, 180, 182
  Bits and Atoms lab 182
Moore, Martin, *Democracy Hacked* 217
Moravec, Hans 195–196
Morley, David 53
Morozov, Evgeny 5, 222
  Net Delusion 222
  To Save Everything, Click Here 232
Mulhern, Francis 140, 143–144, 152–154, 163n.1
Mumford, Lewis 45, 170, 183n.4
Murphie, Andrew 144
Murray, Charles S 83
MySpace 227

NAACP (National Association for the Advancement of Colored People) 115
National Council for Civil Liberties 92
Naughton, John 36
Neville, Richard 83
  Playpower 100n.33
New Scientist 73, 90, 98n.5, 101n.56
New Weird, the 202–204, 206
Nietzsche, Friedrich 43
Noble, Safiya, *Algorithms of Oppression* 231
Norton, Anne 122

O'Neil Cathy, *Weapons of Math Destruction* 231

OOO (Object-Oriented Philosophy) 206
O'Riordan, Kate 48, 188
Ortolano, Guy 142
Orwell, George, *1984* 17, 210
Oxnam, Bishop 81
Oz (magazine) 70, 83–84, 87–89, 99n.29, 100n.3, n.31, n.32, n.34
see also Neville, Richard

Parikka, Jussi 45, 48–50
Pariser, Eli 5
Filter Bubble 233
Peel, John 88, 90, 101n.56
Nightride 88
Perec, Georges 27, 53
Perez, Caroline C., *Invisible Women* 231
Perks, Harry 116
Perlo, Victor 114–116, 132n.14
Philbrick, Herbert 77–78
Phillips, Whitney 229–230
This Is Why We Can't Have Nice Things 230
Photoshop 37, 39, 63n.3, 227
Piercy, Marge 187, 197, 200–202, 207
He/She/It (UK title *Body of Glass*) 187, 197–201, 213n.8
Women on the Edge of Time 197, 213n.9
platform (digital) 10, 18, 21–25 *passim*, 34–35, 52, 57, 219–235 *passim*
Polanyi, Michael 171–172
postdigital, the *see* digital, the
post-human 204, 211, 212n.2, 220
presentism 34–35, 38, 55, 123, 237
primitivism 29, 77, 235
print industry, UK 12
Profumo, John 150, 162

quantum computing 9
Quinn, Zoe 227, 231
Crash Override 231

Rahm, Lina 41, 46
Rajaniemi, Hannu 28, 187, 207
Causal Angel, The 207
Fractal cycle 208–211
Fractal Prince, The 187
Quantum Thief, The 207
rationality-logicality equation 5, 172–173, 176–177, 181
see also Weizenbaum, Joseph
rationality, technocratic 13, 20–21, 27, 94, 108, 124, 140–157 *passim*, 170, 187
see also Leavis, F.R.
Regulation of Investigatory Powers Act UK (2000) 12, 36
Reich, Wilhelm 82
Rentschler, Carrie 96
Report on the American Communist 98n.13
revolution, cybercultural *see* cybercultural
revolution, cybernation *see* cybernation
Ricoeur, Paul 186, 204, 207, 213n.13
Ritchie, Jean 89
Robins, Kevin 130
Rogers, Carl 179
see also therapist
Roszak, Theodore 170
Rushkoff, Douglas, *Team Human* 223

Schanze, Jens, *Plug & Play* 182
Scheff, Sue, *Shame Nation* 230
Schrecker, Ellen 80, 99n.23
science 7, 97, 142–144, 146, 149–150, 157, 160–162, 170–171, 176, 189–190, 192
computer 24, 171, 174–178, 186
history of 43
popular 193

singularity 188–192
social 110
science fiction 28, 112, 143,
    186–203 *passim*, 212,
    222, 227
sciences, the 140, 142, 145
Scott, Laurence, *Four Dimensional
    Human* 225
Searle, John 172, 224
    Chinese room argument 173
Second World War 41, 75
Seeger, Pete 82
Seligman, Ben B. 113
Shannon, Claude 174
Shapin, Steven 97
Shaw, George Bernard, *Pygmalion* 168
Silicon Valley 40, 42, 95, 203,
    222, 225, 229
Silverstone, Roger 53
Silvey, Ted 117
singularity 182, 186–196, 199,
    207–212, 222–223, 227,
    234
    critical theory 189–191
    cognitive theory 190–191
    discourse 4, 187–194, 200, 211
Siri 182
Skågeby, Jörgen 41, 46
Skinner, Burrhus F., *Beyond
    Freedom and Dignity*
    178–180
Smith Act 76
Snow, C.P. 140, 142–151,
    154–162
    Corridors of Power 157
    Rede Lecture 145, 147, 158
    Strangers and Brothers 157
Snowden, Edward 35, 69, 98n.1
social media 18, 35, 85, 141,
    224–225, 229
sociology of media-technological
    innovation 10
Sorgner, Stefan L. *see* del Val,
    Jaime
Sputnik 124
Stallabrass, Julian 226

Steinmueller, Ed 193
Steyerl, Hito 36
Stiegler, Bernard 6, 48, 183n.9
Storr, Will, *Selfie* 225
surveillance 3, 5, 12, 34, 36–38,
    98n.1, 218, 222, 225,
    229
Sutherland, John 83
Suvin, Darko 213n.4

Taft Hartley Act 76, 79, 81
Taplin, Jonathan, *Move Fast and
    Break Things* 222
taxonomy of anti-computing
        15–30 *passim*, 220–221,
        232, 234
    alternative 24–26
    emotional register 24–25
    materials 26
    binary 16–18
    new provisional (eight-part)
        19–24
    of theoretical positions 26–27
Tay (bot) 231
Tegmark, Max, *Life 3.0* 189
temporality, computational 17, 24,
    27, 45–49, 55, 59, 61,
    212
therapist (psycho-) 54, 168, 170,
    175–177, 179–182
    see also ELIZA
therapy (psycho-) 169–170,
    178–180
    Rogerian 168, 179–180, 182
thinking, computational *see*
    computational
Thompson, Denys 152
time-travel 34
    see also temporality,
    computational
Toffler, Alvin, *Future Shock* 232
transhumanism 188–191, 200, 207
Traub, Valerie 14–15, 49, 60–62
Trilling, Lionel 148
Triple Revolution 106, 110,
    112–113, 121
    see also Ad Hoc Committee

Trump, Donald 35, 85, 141, 220, 233
Turing, Alan 9, 198
  test 172
Turkle, Sherry 173, 177, 180, 183n.6, n.10
  Reclaiming Conversation 224
Turner, Fred 42, 95–96
Two Cultures debate 140–146, 148, 150–154, 158–162
  Huxley and Arnold debate 142
  see also Leavis, F.R.; Snow, C.P.

unemployment 20, 109–110, 113–115, 128
Universal Basic Income 113
universal machine 9, 56
utilitarianism 140, 147, 152, 155–156

Vietnam 85, 101n.45
Virilio, Paul 22
  'general accident' theory 22, 221, 227–228
virtual reality 130
viva activa 11, 119, 121, 123–125
  see also Arendt, Hannah
Voss, Georgina 193

Ward, Chris 88–89, 98n.2
Watson, James 161
Web 1.0 227
Web 2.0 24, 227

Weird 202–204
Weizenbaum, Joseph 5, 16, 168–169, 180–182, 193, 224
  Computer Power and Human Reason 169–172, 178, 181
Wendling, Mike, *Alt Right* 229
Wiener, Norbert 9, 108–109, 112, 217
Williams, Raymond 53, 145, 163n.6
Wing, Jeanette 9
Winner, Langdon 17–18
Winthrop, Henry 110, 113
Wired 95–96, 193
work 113–130 *passim*
  end of 105, 109, 116–118, 120, 128
  post-work world 111, 118, 120
Wood, Ellen M. 57–58
Wurman, Richard S., *Information Anxiety* 232

Young-Bruehl, Elisabeth 107, 122–124
YouTube 225, 227

Zielinski, Siegfried 50, 97
Zuboff, Shoshana, *The Age of Surveillance Capitalism* 222
Zuckerberg, Mark 18, 220

EU authorised representative for GPSR:
Easy Access System Europe, Mustamäe tee 50,
10621 Tallinn, Estonia
gpsr.requests@easproject.com

www.ingramcontent.com/pod-product-compliance
Ingram Content Group UK Ltd.
Pitfield, Milton Keynes, MK11 3LW, UK
UKHW021830210426
5322IPUK00004B/121